建筑与市政工程施工现场专业人员继续教育教材

工程施工实用英语

王义新　编著

中国建筑工业出版社

图书在版编目（CIP）数据

工程施工实用英语 / 王义新编著. — 北京 ：中国建筑工业出版社，2023.5

建筑与市政工程施工现场专业人员继续教育教材

ISBN 978-7-112-28491-7

Ⅰ. ①工… Ⅱ. ①王… Ⅲ. ①建筑工程－工程施工－英语－继续教育－教材 Ⅳ. ①TU74

中国国家版本馆 CIP 数据核字(2023)第 051637 号

本书共分为10章，主要内容包括建筑材料、开挖和回填、基础、模板系统、钢筋、混凝土、砂浆、潮湿及其防护、抹灰、地板和吊顶。每个章节包括3个阅读材料、1个对话和1个练习。3个阅读材料是围绕相应章节所讨论的主题而展开的，涵盖相应施工活动的现场和材料准备、工序流程、试验采样和试验流程。对话环节是为了帮助读者进一步熟悉施工现场常用的一些口语词汇和表达方式。填空练习主要是为了帮助读者熟练掌握一些关键词汇并提高相应的阅读能力。本书可供在校土建专业的学生，以及从事国际工程项目的工程技术人员和现场翻译人员等参考使用。

责任编辑：葛又畅
责任校对：孙　莹

建筑与市政工程施工现场专业人员继续教育教材
工程施工实用英语
王义新　编著
*
中国建筑工业出版社出版、发行（北京海淀三里河路9号）
各地新华书店、建筑书店经销
北京红光制版公司制版
建工社（河北）印刷有限公司印刷
*
开本：787毫米×1092毫米　1/16　印张：8½　字数：207千字
2023年5月第一版　2023年5月第一次印刷
定价：**33.00**元
ISBN 978-7-112-28491-7
(40952)

前　言

改革开放以来，尤其随着“一带一路”倡议的深入落实，我国建筑企业参与的国际工程施工项目越来越多，这对人才的培养提出了更高的要求，培养一批精通一门或者多门外语的工程技术人员，满足国际工程施工项目对人才的需求显得尤为重要。本教材正是为了适应和满足这一需求而编写的。

本教材共分为10章，主要内容包括建筑材料、开挖和回填、基础、模板系统、钢筋、混凝土、砂浆、潮湿及其防护、抹灰、地板和吊顶。

建筑材料是施工活动的基础，因此将其作为教材第1章的内容。其余章节的先后顺序是参照土建项目施工活动的一般流程安排的：土方工程中的开挖和回填、地基与基础工程中的基础分别作为第2、3章；钢筋混凝土工程中的模板系统、钢筋和混凝土这3个章节安排在教材的核心位置；砌体工程中的砂浆在内容方面与混凝土存在一些相近之处，故将其安排在混凝土之后；屋面工程中的潮湿问题及其防护和装饰工程中的抹灰、地板和吊顶作为教材的最后3个章节。

每个章节包括3个阅读材料、1个对话和1个练习。3个阅读材料是围绕相应章节所讨论的主题而展开的，涵盖相应施工活动的现场和材料准备、工序流程、试验采样和试验流程。对话环节是为了帮助读者进一步熟悉施工现场常用的一些口语词汇和表达方式。填空练习主要是为了帮助读者熟练掌握一些关键词汇并提高相应的阅读能力。

本教材体现3个特点：(1) 合理性，章节的结构以土建项目施工的顺序为主线进行安排；(2) 实用性，章节的阅读材料和对话紧紧围绕施工活动的实务而展开；(3) 针对性，主要针对在校土建专业的学生，以及从事国际工程项目的工程技术人员和现场翻译人员等。

本教材在编写过程中，参考了大量相关的教材和资料，在此向这些教材和资料的作者表示感谢。

需要说明的是，在本教材的修改、校核过程中，得到了中国建筑工业出版社编辑朱首明、葛又畅，以及新西兰友人 Tim Jelleman 的无私帮助和指导；浙江国兴投资集团有限公司的工程师寿泽恩参与了全书的审稿工作，对本教材的中文内容从专业方面提出了大量的宝贵建议；同济大学的王晨晓对本教材的中文内容进行了整体的文字校正和润色；同济大学土木工程学院郑朝晖和陈建翔帮助审核了清样稿。在此向他们表示最诚挚的谢意！

鉴于作者的水平及现场经验有限，书中难免存在不足之处，敬请读者批评指正。

目　录

Unit 1　Building Materials (建筑材料) …… 1

1. Reading：Structural Steel (型钢) …… 3
2. Reading：Cement (水泥) …… 5
3. Reading：Mortar (砂浆) …… 7
4. Dialogue：Delivering Rebar on Site (钢筋验收) …… 9
5. Exercises (练习) …… 11

Unit 2　Excavation and Backfilling (开挖和回填) …… 13

1. Reading：Buried Services (预埋管线) …… 15
2. Reading：Excavation (开挖) …… 17
3. Reading：Backfilling and Compaction (回填和夯实) …… 20
4. Dialogue：Field Density Test (密实度试验) …… 22
5. Exercises (练习) …… 23

Unit 3　Foundation (基础) …… 25

1. Reading：Types of Foundation (基础类型) …… 27
2. Reading：Footing Near Slope (边坡基础) …… 29
3. Reading：Dewatering (降水) …… 32
4. Dialogue：Boring Pile Wall (钻孔灌注桩) …… 35
5. Exercises (练习) …… 37

Unit 4　Formwork (模板系统) …… 39

1. Reading：Formwork and Its Materials (模板系统和材料) …… 41
2. Reading：Types of Formwork (模板系统类型) …… 43
3. Reading：Construction of Formwork (模板系统搭建) …… 46
4. Dialogue：Scaffolding Erection (脚手架搭建) …… 48
5. Exercises (练习) …… 50

Unit 5　Rebar (钢筋) …… 51

1. Reading：Rebar Test (钢筋试验) …… 53
2. Reading：Placement of Rebar (钢筋铺设) …… 54
3. Reading：Rebar Ties and Supports (钢筋支撑绑扎) …… 57
4. Dialogue：Tying Rebar (钢筋绑扎) …… 60
5. Exercises (练习) …… 62

Unit 6　Concrete (混凝土) …… 63

1. Reading：Concrete Manufacturing (混凝土制备) …… 65
2. Reading：Concrete Handling and Placing (混凝土运输与泵送) …… 67

3. Reading：Concrete Curing（混凝土养护） …… 70
4. Dialogue：Concrete Slump Test（混凝土坍落度试验） …… 71
5. Exercises（练习） …… 74
Unit 7 Mortar（砂浆） …… 77
1. Reading：Mortar Ratio（砂浆混合比） …… 79
2. Reading：Methods of Mixing Mortar（砂浆搅拌方式） …… 80
3. Reading：Brick Laying（砌块） …… 83
4. Dialogue：Procedures of Mixing Mortar（砂浆搅拌工序） …… 86
5. Exercises（练习） …… 88
Unit 8 Dampness and Its Prevention（潮湿及其防护） …… 89
1. Reading：Causes of Dampness（潮湿的原因） …… 91
2. Reading：Dampness Damage and Its Prevention（潮湿的危害和预防） …… 93
3. Reading：Dampness Proofing（防潮） …… 95
4. Dialogue：Waterproofing in Toilet（卫生间防潮处理） …… 97
5. Exercises（练习） …… 98
Unit 9 Plastering（抹灰） …… 101
1. Reading：Wall Plastering Preparation（抹灰准备） …… 103
2. Reading：Procedures of Mixing Plaster（泥浆搅拌） …… 104
3. Reading：Wall Plastering 1（墙面抹灰 1） …… 106
4. Dialogue：Wall Plastering 2（墙面抹灰 2） …… 108
5. Exercises（练习） …… 111
Unit 10 Flooring and PVC Ceiling（地板和吊顶） …… 113
1. Reading：Vinyl Flooring Installing（PVC 地板的安装） …… 115
2. Reading：Tile Flooring（瓷砖地板） …… 117
3. Reading：PVC Ceiling Installing（PVC 吊顶安装） …… 121
4. Dialogue：Procedures of Installing Vinyl Flooring（PVC 地板的安装工序） …… 124
5. Exercises（练习） …… 126
Key to Exercises（答案） …… 127
参考文献 …… 128

Unit 1

Building Materials （建筑材料）

1. Reading: Structural Steel (型钢)

(1) Introduction

Structural steel is very popular in construction, here its shapes, application and two types of connection will be discussed.

(2) Text

Steel is well known for providing structure and strength unlike any other when it comes to construction. Also, the durability and versatility that steel provides are not matched by either wood or concrete. In addition, steel construction has many other advantages: the strength-to-weight ratio is excellent, steel joins easily, efficient shapes are available, etc.

Structural steel is a category of steel construction material which is produced with a particular cross-section or shape, and some specified values of strength and chemical composition. Here we take a look at the various shapes of structural steel and what purposes they serve in the process of construction.

Structural steel comes in various shapes like I-beam, L-beam, Z-shape, L-shape (angle), channel (C-beam, cross-section), HSS-shape, T-shape and much more. I-beam or H-beam, is used in construction and civil engineering projects, while angle beam in residential construction, mining and infrastructure; channel in the construction of bridge; hollow steel section for welded steel frame. The reason why structural steel is so popular in construction is due to its many advantages.

One of the structural steel advantages is machinability which allows builders to weld or bolt the materials into a variety of shapes. So, there are two types of connection: pin connection by bolting, and fixed connection through welding. A pin connection is created by bolting a beam to a column with clip angles along with the beam web. This means that the beam should not be able to move up or down, nor in or out, but it can rotate a bit. On the contrary, the fixed connection must stop that ability to rotate since the beam and column are welded together. Connections are very important with structural steel.

Another advantage of structural steel is fire-resistant in itself but fire protection methods should be put in place in case there is a possibility of it getting heated up to a point where it starts to lose its durability and strength. Furthermore, corrosion must be prevented.

To take advantage of structural steel, such tests as yield and tensile strength and Charpy impact should be conducted: yield and tensile strength help to determine the steel grade and overall application, and the Charpy impact test result tells engineers if the material is adequate for their project.

(3) Words and Phrases

durability　耐久性，耐用性
versatility　多面性，多功能
composition　组成，构成，成分
hollow steel section　空心型钢
machinability　机加工性能
clip angle　扣角钢
yield strength　屈服强度
tensile strength　抗拉强度
Charpy impact test　摆锤式冲击试验

译文

(1) 导读

型钢在建筑施工中应用得非常广泛。在这里，本篇将着重介绍它的形状、在建筑施工中的应用，以及两种连接方式。

(2) 正文

众所周知，在建筑行业中，钢材在结构和强度方面的性能远远优于其他材料。同时，钢材的耐久性、适用性也比木材或者混凝土这些材料高出很多。此外，钢结构还有许多其他优点：比强度优异、材料连接容易、形状多样等。

型钢是一种有特定截面或形状、有一定的强度值和化学组成的钢材。型钢的各种形状种类及其在建筑活动中的应用将是本文的重点。

市场上有不同形状的型钢材料，例如工字钢、角钢、Z型钢、L型钢（角钢）、槽钢（C梁）、HSS型空心型钢和T型钢等。工字钢或H型钢，通常使用在建筑和土建工程项目中；角钢用在民用建筑、矿山和基础设施建筑中；槽钢用在桥梁建设中；空心型钢用来焊接钢框架。型钢在建筑中被广泛应用的原因是它自身具有许多不同的特性。

型钢的其中一个特性就是它的机加工性能良好，它能通过焊接或螺栓连接组合形成各

种不同形状。这里通常有两种类型的连接方式：使用螺栓形成的销连接和通过焊接形成的固定连接。把梁和柱用螺栓进行销连接是指沿着梁腹板用扣角钢来进行连接，在这种连接方式下，梁不能发生上下方向的移动或进出，但能够稍微转动一下。与之相反的是固定连接方式，此方式下梁不能进行任何的旋转，因为梁和柱子是焊接在一起的。对于型钢来说，连接方式非常重要。

型钢的另外一个特性是它自身的耐火性能优异，但是防火措施也应该做到位，以防它可能因受热过度而导致失去自身的耐久性和强度。此外，还应该对型钢自身进行相应的防锈处理。

为了能够充分发挥型钢的各种特性，应该进行各种相关的试验，例如屈服强度试验、抗拉强度试验和摆锤式冲击试验。屈服强度试验和抗拉强度试验可以确定钢材的规格等级和总体适用性，而摆锤式冲击试验的结果会告诉人们该型钢材料是否适用于某一特定项目。

2. Reading：Cement （水泥）

（1）Introduction

Cement is one of the most commonly used materials on construction sites. Its features，types，and application are the knowledge engineers must be familiar with，therefore they are discussed as follows.

（2）Text

Cement is a kind of binding material commonly used on construction sites，which is manufactured in a factory where calcium and clay materials are mixed and burnt at an extremely high temperature to produce clinker which is ground into a final fine power，producing cement.

Normally，types of cement include ordinary Portland cement，white cement，colored cement，quick setting cement，rapid hardening cement，low heat cement，expanding cement，high alumina cement，blast furnace cement，acid-resistant cement，sulfate resistant cement，and fly ash blended cement.

As their type names suggest，different kinds of cement could be used for quite different civil works. For instance，expanding cement could be used to fill the cracks in concrete structures，while low heat cement in mass concrete works like dams.

As for properties of ordinary Portland cement，on one hand，cement itself is composed of tricalcium silicate, dicalcium silicate, tricalcium aluminate, and tetracalcium aluminate. Any of these would react differently after mixing with water. On the other hand, cement has five physical properties, including fineness, setting time, soundness, compressive strength and tensile strength, which is needed to be tested to show how the cement will behave after adding water.

After all, cement is so common that it is used for repairing cracks in structures, producing lamp posts, communication posts, garden seats, cement pipes, and railway sleepers, serving as materials for buildings, bridges, tunnels, roads, footpaths, and the like.

(3) Words and Phrases

cement　水泥

clay　黏土

clinker　熔块，熟料

Portland cement　普通矿渣硅酸盐水泥

quick setting cement　快凝水泥

flash ash　飞灰

fineness　细度

setting time　凝结时间

soundness　安定性

crushing strength　抗压强度

(1) 导读

水泥是工程施工现场上最常用的材料之一，因此了解它的属性、种类和应用对工程人员至关重要。本篇将着重介绍一下水泥的生产过程、种类、属性和应用场合。

(2) 正文

水泥是一种在施工现场上经常使用的粘合材料，它的原材料是石灰石（氧化钙）和黏土。在水泥制造厂里，工人首先把这两种原材料混合起来，接着在高温下把它们烧结形成水泥熟料，最后将这些水泥熟料研磨成极细的粉状物，也就是水泥成品。

一般来讲，水泥的种类包括普通矿渣硅酸盐水泥、白水泥、有色水泥、快凝水泥、速硬水泥、低热水泥、膨胀水泥、高矾土水泥、高炉水泥、耐酸水泥、抗硫酸盐水泥和飞灰混合水泥。

正如上述名称所示，不同类型的水泥适用于截然不同的土建工程项目中。例如，膨胀水泥通常用于修补混凝土结构浇筑过程中出现的缝隙；低热水泥通常使用在大型混凝土施工项目中，比如水利大坝的建设。

对于普通矿渣硅酸盐水泥来讲，一方面，水泥自身由三钙硅酸盐、硅酸二钙、三钙铝酸盐、四钙铝酸盐构成，这些成分在与水混合后产生的反应也各不相同；另一方面，水泥还有五种物理属性，分别是细度、凝结时间、安定性、抗压强度和抗折强度，需要对水泥的这些物理属性进行检验试验，以此来确定其在以后的使用过程中的性能表现。

总的来讲，水泥的使用场合非常广泛。例如，水泥可用于修补混凝土结构浇筑过程中所出现的缝隙，同时可应用于路灯水泥柱、通信线路杆、公园座椅、水泥管道、铁路道枕、房屋、桥梁、隧道、道路、人行道以及其他构件的浇筑和制造中。

3. Reading：Mortar （砂浆）

(1) Introduction

Mortar is another one of the most common materials used on construction sites. How to mix various ingredients thoroughly in a correct ratio affects its performance, this issue will be discussed as follows.

(2) Text

Mortar is a mixture of binding material, fine aggregate, and water. It is produced by adding water into a mixture of binding material and fine aggregate. Among the ingredients of mortar are, first, binding material which serves to join sand and other construction materials of bricks and stones together, as well as fill the voids and joints in structures; second, fine aggregate which functions to form a thin film of mortar paste that is easy to spread and adhere to other materials, as well as fill the joints of bricks and stones.

In terms of the types of mortar, they could be cement mortar, lime mortar, and mud mortar due to the type of binding materials, like cement and lime.

For cement mortar, it is formed by mixing sand and cement thoroughly in a dry condition, either by hand or by machine, then adding water in a small amount at one time, next stirring or shoveling them upside down for sufficient times, until the consistency feels right.

Remember that the correct proportion of cement and sand is vital for its strength, which requires adequate time to set. Generally speaking, cement mortar should be kept wet for some time to develop strength either by watering or spraying curing compound or being covered by plastic sheets. Cement mortar is very popular on construction sites and used to bind stones, bricks, and cement blocks, to plaster slabs or walls, as well as other civil works.

For lime mortar, it could be formed by adding stone-lime through pounding and grinding, depending on the quantity of mortar needed.

For mud mortar, it is produced by mixing clay lumps and water which is less used in modern construction sites.

What's more, there are some tests, such as the water retention, compressive strength, tensile strength, and adhesive test, which could and should be done to show the suitability of mortar for a specific work.

(3) Words and Phrases

inert material　惰性材料
ingredient　组成成分
spread　摊，铺
adhere　粘附
lime　石膏
mud　泥浆，污泥
stir　搅拌
shovel　铲
curing compound　养护剂
pounding　敲打
grinding　粉碎

(1) 导读

砂浆是施工现场上经常使用的另外一种建筑材料，其混合比正确与否以及搅拌均匀程度的高低直接关系到砂浆的各项工作性能。本篇将着重介绍砂浆的组成成分、混合和搅拌的方法等基本知识。

（2）正文

砂浆是一种由胶凝材料、细骨料和水组成的混合物，它是通过把水和由胶凝材料、细骨料形成的混合物进行搅拌而形成的。在砂浆的组成成分中，胶凝材料用来将细骨料和其他建筑材料例如砖石粘结起来，以及填充建筑结构件里的缝隙和孔洞；细骨料被用于形成一层薄薄的砂浆来抹敷粘贴在其他材料上，以及填充砖石结构中的缝隙。

砂浆的类型包括：水泥砂浆、石灰砂浆和黏土砂浆，它们的区别在于胶凝材料的不同，三者的胶凝材料分别为水泥、石灰膏和黏土。

对于水泥砂浆来讲，首先在干燥的环境下将细骨料和水泥混合均匀，可以通过人工完成也可以通过机器进行搅拌，然后向其中加水，应采用多次加少量水的方式，接着通过机器搅拌或人工用铲上下翻铲，使之均匀，注意翻铲的时间要足够，当黏稠度达到均匀一致时停止。

需要特别注意的是，水泥、细骨料的混合比例将关系到砂浆的最终强度，同时砂浆需要适当的时间来凝固。一般来讲，水泥砂浆需要按照规定保湿一段时间，保湿方法可以是洒水，也可以是喷洒养护剂，或者是覆盖塑料膜。施工现场上水泥砂浆的应用非常广泛，可用于砖石和混凝土砌块的粘结、楼板或墙体的找平，以及其他土建工程中的施工活动。

对于石灰砂浆来讲，石灰石可以经过敲碎或者粉碎处理之后，再进行搅拌混合，具体采用哪一种方式应根据所需要的砂浆量来决定。

至于黏土砂浆，它是通过将黏土块和水混合起来而形成的砂浆材料，在如今的施工现场上黏土砂浆已经很少使用了。

此外，对于砂浆通常还要进行一些试验，例如保水性试验、抗压强度试验、抗拉强度试验和粘附性试验，来进一步验证某一特定批次的砂浆是否满足指定项目的质量要求。

4. Dialogue：Delivering Rebar on Site （钢筋验收）

At 9 a. m. on Monday，on one construction site，Jack and David are in the rebar yard，receiving one truck of rebar. Jack is showing a rebar shipping ticket to David，an intern.

Jack：Here is a rebar ticket，from one of our suppliers.

David：One truck one ticket. Is that right?

Jack：Absolutely. We learn these rebars from it.

David：What if the ticket and the rebars do not match?

Jack：We will call the supplier. If the rebars are not ours，we get them taken back.

David：What will we do if they are our rebars?

Jack：We will take some samples from this truck of rebar.

David：Do we do it on ourselves?

Jack：No way！John，the engineer from consultancy，will determine how to sample for testing and our labors will cut accordingly.

David：Where is the lab for testing them?

Jack：The testing lab is in the downtown area. We go there together with John.

David：Do we get the result on the same day as we go there?

Jack：Definitely！We wait till the result comes up，probably in two hours.

David：What comes next if the result is bad?

Jack：We will ring the supplier and let them replace rebars with new ones.

David：If it is good?

Jack：If that is the case，rebars will be bent and cut according to the rebar schedule. Look！John is here. Let us meet him.

周一上午9点，在一个施工现场中，杰克和大卫在钢筋加工区域接收了一卡车的钢筋。大卫是一名实习生，杰克正在向他解释钢筋的发货单。

杰克：这是这一卡车钢筋的发货单，是我们的一个供应商提供的。

大卫：是一车一个单据吗？

杰克：是的，本批钢筋的性能指标都在上面写着呢。

大卫：假如发货单和钢筋对不上的话，我们该怎么办？

杰克：我们会给供应商打电话来核实，若是发错货的话，他们会主动拉回去的。

大卫：那如果钢筋是我们订购的呢？

杰克：我们会从这车钢筋中截取一些试验样品。

大卫：我们自己取样吗？

杰克：不行的，约翰来决定取样的数量，他是咨询公司的一名工程师，我们的工人应该依据他的要求来截取试样。

大卫：那检测钢筋试样的试验室在哪里？

杰克：试验室在市中心，到时候我们和约翰一起去。

大卫：当天我们能得到试验结果吗？

杰克：当然，我们会在那里等结果出来，一般2h。

大卫：如果试验结果不好的话，我们该怎么办？

杰克：那样的话，我们只能打电话给供应商，让他们再送批新的钢筋来。

大卫：那结果如果是合格的话呢？

杰克：如果试验结果没有什么问题，这些钢筋就会按照加工图纸的要求，进行相应的弯曲和切割。你瞧，约翰到了，我们过去迎接他一下。

5. Exercises （练习）

(1) Mix design of concrete is a process of calculating the quantity of materials like cement，sand， 1 ，water，and admixtures to achieve or make specified strength of concrete.

(2) Concrete has a good 2 strength but a poor tensile strength，that is the reason why the reinforcing steel is needed and placed before pouring the concrete.

(3) There are two kinds of wood：hardwoods and 3 . They are used as main building materials.

(4) Plywood is sheet construction material，made from odd numbers of thin 4 of wood veneer.

(5) Welded wire mesh is a kind of welded fabricated grid，made of low 5 steel wire or stainless-steel wire.

Unit 2

Excavation and Backfilling

（开挖和回填）

1. Reading：Buried Services （预埋管线）

（1）Introduction

It is common to encounter some buried services during excavation. Those buried cables or pipes must be protected carefully，or they could be damaged and out of service. What should be done with them before and during excavation will be discussed below.

（2）Text

Local utilities and other services providers should be contacted before commencing excavation to get information on the size，depth，and location of buried services which include gas，electricity，water，communication，and other lines or cables. If information is not available or difficult to get，some careful soil exploration should be done to see if there are any buried lines.

During excavation，representatives from affected utilities should be notified beforehand and be present on time to advise on how to protect those lines or cables；when lines are damaged，immediate repair should be done correctly under the supervision of representatives from utilities.

There may also be some obsolete buried lines or structures which might be encountered during excavation，these lines or structures should be cut off or broken down into small pieces before being dumped in the designated areas outside the excavation site.

In brief，buried services，whether currently active or obsolete，require special attention before and during excavation.

In and around the excavation site，there may be some trees，lawns，utility poles，fences，pavement，and other buildings which are needed for current and future use. These existing objects or structures should be protected during excavation by barricading，shoring，suspending，or by other means，as required. In event of any damage，timely repairs should be carried out to restore them to the status as they were before.

（3）Words and Phrases

utility 供电部门

excavation 开挖

soil exploration 地下探测

buried line　预埋管线
cut off　切割
dump　倾倒
barricade　栅栏
shoring　支撑
suspending　悬挂

（1）导读

在开挖过程中，人们可能会遇到一些预埋在地下的市政管线，这些管线应该在开挖的过程中保护好，以免其被损坏或者影响市政供电、供气等业务的正常运行。本篇将着重介绍开挖之前和过程中应该注意的一些事项。

（2）正文

在正式开挖之前，应该联系当地的市政部门，搜集埋设在施工现场附近的市政线路的敷设位置、深度、管线尺寸大小等信息。市政线路包括燃气、电力、自来水、通信及其他地下敷设的市政管线或电缆。如果无法获得市政管线敷设的相关技术资料，则应该在施工现场安排进行一些地下钻探活动，来确定是否存在以前预埋的市政管线或电缆线路。

应该确保有关市政部门能够提前接到开挖通知，并且他们的人员代表能够及时到达现场，以便在开挖过程中针对如何保护市政管线提供相应的技术指导。在开挖过程中，市政管线一旦遭到破坏，则应当在市政人员的监督下立即对其进行修复，使之恢复完好。

在开挖过程中，也可能遇到一些埋设在地下的已经报废或过期的市政管线或结构体，这些市政管线或结构体应该被切割成更小的建筑垃圾，然后运输到开挖现场以外的指定堆放区域。

无论是在开挖之前还是在开挖过程中，对于正在运营的或者已经报废过时的预埋市政管线，现场施工人员都应该特别重视。

另外，在开挖现场或其附近区域还可能会有一些草坪、树木、电力通信杆架、栅栏或其他建筑物，它们在当前或者以后都还具有使用价值，因此在开挖过程中应该对这些设施或结构体采取适当的保护措施，如隔离、支撑、悬挂或其他必要的形式。如果它们有任何的损坏情况，应该立即组织人力对其进行维修，使之修复到最初的完好状态。

2. Reading：Excavation（开挖）

（1）Introduction

Excavation is the first and one of the hugest jobs in a construction project，involving the safety of properties，utilities，and people working in and around its site. Here the types of excavation，precautions to be taken before excavation，and protection system to be put in place during excavation will be discussed.

（2）Text

Excavation is the process of cutting，digging，and removing the earth below the ground level to the desired levels（as in Fig. 2-1），grades and elevations，and produces cavity，trench，and depression on the earth's surface.

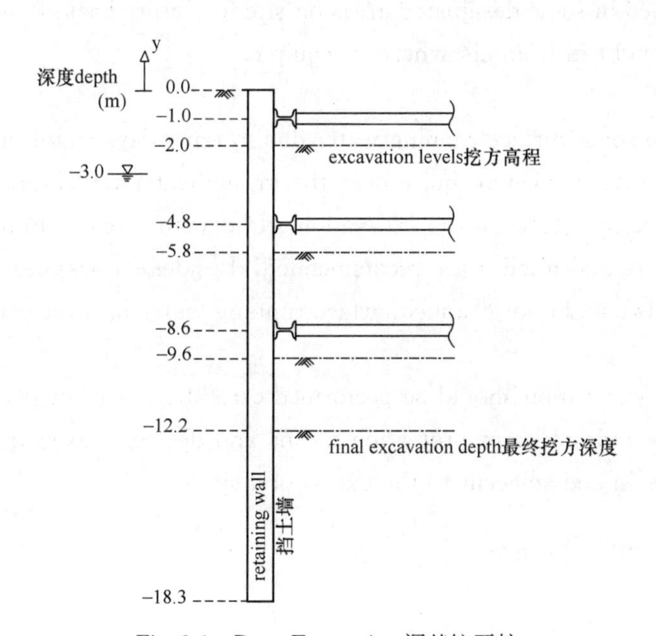

Fig. 2-1 Deep Excavation 深基坑开挖

It is classified into common excavation and unauthorized excavation. Common excavation involves traditional means of digging，cutting，trenching，sloping，shoring，etc. For unauthorized excavation，since it is unwanted，some correction should be done according to the instruction from engineers.

As required by construction plan，markings and stakes for excavation will have already been laid by surveyors. Excavate to the desired levels，grades，and elevations with the right machines，while establishing a proper protection system like sloping，shoring， trench box, shields, and others, if necessary, when excavation progresses.

When excavation is in process, regular inspection on the elevations of excavation should be done by competent personnel to make sure that over-excavation is not done. If over excavated, some corrective measures should be approved by authorized engineers and performed under supervision. Those corrective measures could include placing some suitable materials and compacting them according to the requirement of codes.

During excavation, some abandoned underground service lines and structures should be broken, cut into small pieces, and removed, apart from capping some pipes with gas and water. Those underground objects may include sewers, water lines, gas lines, utility cables, and others. Therefore, materials produced from excavation might be either unsuitable materials or suitable materials. On one hand, unsuitable materials should be transported to a designated area outside the project site. On the other hand, suitable materials should be stockpiled in some designated areas on site for future backfilling, combined with some new soil brought in from elsewhere if required.

There may be some surface running water due to rainy days and groundwater seepage if the elevation bottom is near or quite near the groundwater table. Hence pumping and dewatering are necessary to keep the excavation site dry and free from mud. Sometimes, water collection pits and ditches are recommended. Besides, stockpiles of excavated soil and other materials should not channel surface running water into the excavation site.

In summary, excavation should be performed carefully not to be over excavated, and safely with proper and sufficient protection means and devices, apart from not damaging existing structures in and adjacent to the excavation site.

(3) Words and Phrases

dig 挖掘

elevation 标高，高程

cavity 空洞

trench 沟渠

depression 凹陷

unauthorized excavation 超挖

stake 木桩

capping 封堵

dewatering 降水

译文

（1）导读

开挖是土建施工中的第一项也是工程量最大的一项施工活动，它同时涉及现场的设备资产、公用设施以及现场内和周边人员的安全问题。本篇将着重介绍开挖方式、开挖前应该采取的预防措施和开挖过程中的一些保护措施。

（2）正文

开挖是一项施工过程，它通过挖掘和清除地表面以下的土壤到设计要求所规定的标高（图 2-1）、台面和坡度，来形成孔洞、沟渠和基坑。

开挖分为常规开挖和超挖。对常规开挖来讲，它采用的是一些传统的开挖方法，比如挖掘、挖沟、放坡式开挖、挡板支撑开挖等。对于超挖来讲，应该根据工程设计人员的技术要求进行必要的纠正施工。

在开挖施工之前，测量员应该已经按照设计图纸对开挖区域进行了画线和打桩工作。在开挖过程中，首先应保证采取了及时的、适当的保护措施，比如进行了放坡、支撑、安装地连墙、地盾保护以及其他必要的保护作业；其次再使用适宜的设备开挖到设计要求的标高、台面和坡度。

在开挖过程中，应该由有能力的工程监理人员定期对开挖的标高进行检查核验，从而防止超挖的情况出现。一旦出现超挖，应采取纠正措施，施工方案必须经过授权工程人员的批准，同时施工应在他们的有效监督下进行。纠正措施可能是回填一些适宜的土石材料，并根据规范要求进行夯实。

在开挖过程中，应该对遇到的废弃地下管线和建筑结构体进行必要的、适当的处置，如将其切割成更小的建筑垃圾并清除出现场，或原地封堵住燃气和自来水管线的出口。那些地下管线可能是排污管道、自来水管道、燃气管线、电力电缆或其他设施。因此，在开挖过程可能会产生一些建筑垃圾，也可能产生一些未来可被再次利用的回填材料。一方面，建筑垃圾应该被运送至施工现场以外的指定堆放区域；另一方面，回填材料应该在现场的指定堆放区域内堆放，以备将来回填使用，若有特殊要求，回填材料也可与从别的地方开挖的新土混合在一起使用。

此外，开挖现场经常会出现积水，一种可能是下雨形成积水，另一种可能是开挖的深度接近地下水位，导致地下水渗透而形成积水。因此，有必要利用水泵向外排水，并对地下水进行必要的降水处理，以保证施工现场干燥且不出现污泥，有时还可以在开挖现场设置一定数量的集水井和排水沟。另外，开挖出来的土方及其他施工材料的堆放也应符合相应的规范要求，确保地面上的雨水不会流入开挖现场。

总之，开挖施工应该谨慎进行，以防止出现超挖的情况，同时要及时采取安全保护措施并使用相关的设备来保证施工的安全，并且不能对施工现场及周围的设施造成破坏。

3. Reading：Backfilling and Compaction（回填和夯实）

(1) Introduction

Backfilling and compaction are the last steps of earthing work, also two of the most important jobs regarding excavation. Here some precautions needed to be taken beforehand and procedures of compaction will be discussed as follows.

(2) Text

1) Backfilling

Structure backfilling should use suitable soil excavated from the site and new borrowed soil from an area designated by engineers if necessary. This soil should be placed and spread in layers by backhoe or hand tools. Before placing soil, all construction debris, grasses, roots, and other foreign materials should be removed. Be reminded that placing and compacting the soil should be done in a "dry" condition, which means no running water or groundwater is present on spot. If there is surface running water, pump or channel them out.

2) Compaction

Firstly, choose the right and proper compaction machines based on the types of soil and constraints of site. Secondly, make sure there is no frozen material or ice in the soil. Then, make test on the moisture content of the soil to see if it meets the requirement of the concerning codes and regulations. Spray sufficient water on the soil if the soil is too dry, or let the soil dry for some time by exposing it to the sun if the soil is too wet. Next, spread and level the soil in layers of 300mm, before compacting them with heavy vibrator tampers or hand tampers to the desired density according to the specification and codes. After that, perform the field density test on the compacted layers of soil. Finally, submit the field density test report to the authorized engineers and proceed with the next work, otherwise compact again, or add more other suitable materials before compacting again, till the density of all layers satisfies the requirement of codes.

When installing pipes in trenches, place granular materials and compact them well as a bedding to support pipes.

(3) Words and Phrases

backfill　回填

borrowed soil 外采挖土方

layer 回填土层

backhoe 反铲挖土机

debris 残渣

compact 夯实

running water 地上流水

granular material 粒状材料

moisture content 含水量

vibrator tamper 震动式夯土机

field density test 密实度试验

（1）导读

回填和夯实是土方施工过程中的最后两项工作，同时也是最重要的两项任务。本篇将着重介绍夯实前应该做的一些准备工作以及夯实的一些步骤。

（2）正文

1）回填

混凝土结构的回填应该利用从开挖现场挖掘出来的回填土方，如果有必要的话，也可以从工程监理人员指定的区域采挖新的土方。这些土方需要用反铲挖土机或手工工具一层一层地摊铺，摊铺之前必须清除干净所有的建筑垃圾、草木、树根及其他杂物。这里需要注意的是，土方的摊铺夯实均需在干燥的环境下进行，也就是说，在回填现场不能有积水或者渗出的地下水。若回填现场发现有积水，则需要使用水泵把水抽出去或者修设水槽把水引出现场。

2）夯实

首先，需要根据土方的种类和现场的情况选用合适的夯实机械。其次，要确保土方中没有冻土或冰存在。然后，对要夯实的土方进行取样并进行含水率试验，以确认其是否满足相应标准规范的要求。如果土方过于干燥，那么需要在土方上喷洒适量的水；如果土方过于潮湿，那么需要让它在太阳下暴晒一段时间，使其含水率达标。接着，一层一层地摊铺土方并使其平整，使每层土方的厚度为300mm，再用大型夯实设备或者人工来夯实，以使土方的密实度能够符合相关标准规范的要求。随后，对夯实过的每层土方进行取样，并检测其密实度。最后，向经过授权的工程监理人员提交现场密实度试验报告，若试验合格方可进行后续的相关施工；若密实度试验的数据不达标，则需要重新夯实，或者先回填一些砂土再进行夯实，直至每层的密实度数据符合规范要求再进行后续施工。

若需要在沟渠中安装管道，则需要事先在其底部铺上一些砂石材料，并将铺上的砂石材料夯实到相应的强度，使其作为支撑管道的垫层。

4. Dialogue：Field Density Test （密实度试验）

At 10 a. m. on Tuesday，on one construction site，workers are backfilling and compacting the foundation of one concrete structure. John，the consultant，is taking samples from the soil with help from David，an intern，while Jack is explaining field density test to him.

Jack：We will take three samples from this section，and this is the first layer.

David：When are we going to take them to a lab for test?

Jack：Probably in the afternoon. John will drive there；we go with him.

David：When will the report be due?

Jack：The day after tomorrow，I guess. John will go and collect it.

David：If the result is not OK，what is next?

Jack：Compact again，or bring some new soil before compacting，if necessary. We will see how it goes.

David：Proceed to backfill the second layer of soil，if the result is fine，am I right?

Jack：Certainly，some materials for backfilling will be brought in，spread，and leveled.

David：How thick must each layer be at least?

Jack：At least 300 mm.

John：All right，we are done with the first sampling. Jack，could you please take this sample to my car? I go over there for a second sampling with David right now.

Jack：Of Course，I catch you guys a little while later.

周二上午10点，在一个施工现场中，施工工人正在对一个混凝土结构的基础进行回填和夯实。约翰是咨询公司的一名工程师，大卫是一名实习生，约翰正在大卫的协助下对夯实土层进行取样，而杰克在旁边给大卫解释密实度试验的相关事宜。

杰克：我们需要从这一片夯实过的土层中取三个试样，这是第一层土。

大卫：我们要在什么时间把这些试样送到试验室做试验?

杰克：可能要等到下午了。约翰开车去，我们可以和他一起去。

大卫：什么时间能出结果？

杰克：大概在后天，约翰会过去取的。

大卫：结果要是不合格的话，我们该怎么办？

杰克：那么需要重新进行夯实，还有可能会在夯实之前再铺上一些回填土。到时候我们看情况再说。

大卫：如果试验结果没有问题，那么就接着回填第二层土，对吗？

杰克：是的，我们会安排再运来一些回填土，接着把它们摊平铺开。

大卫：每个土层的厚度至少是多少？

杰克：至少 300mm。

约翰：好了，我们第一个试样已经提取完毕了。杰克，你能把这个试样放到我的车里吗？我和大卫现在去那边取第二个试样。

杰克：没问题，咱们一会儿见。

5. Exercises （练习）

(1) Groundwater is the water in the soil in which the content of water varies according to the depth of soil and is shown by water __1__.

(2) The most dangerous and common accident faced by excavation workers is __2__ which could bury them suddenly and injure or kill them while they are working.

(3) Before excavation, utilities should be notified so that the services __3__ likely to be affected should be de-energized.

(4) Excavation is followed by other earthworks, like backfilling with the new or original soil, spreading, grading, leveling, and __4__ by earthmoving plants.

(5) When the excavation needs to go down deeper, it is preferable to use __5__ to do the job on behave of human beings due to the risk of soil collapse and boiling of groundwater.

Unit 3

Foundation （基础）

1. Reading：Types of Foundation（基础类型）

（1）Introduction

As well known，the foundation is one extremely important element in construction structure，its quality has something to do with the quality of a whole structure. Here its definition，function，and types will be explained as follows.

（2）Text

Foundation is the lower part of a building，also known as its substructure. Building a foundation is the process of digging the ground into a desired shape and depth，and then constructing a mass body，which is an important part of the construction project and the first one with which all other activities can start. Its function is to distribute and transfer the loads from the superstructure into the soil and resist any forces，including lateral forces of the building itself or wind force，and prevent them from moving and overturning the building. It provides a level surface on which all other structural elements may be built and supported.

Foundations fall into two types：shallow foundations and deep foundations. A shallow foundation is divided into isolated footing and combined footing while a deep foundation into pile foundation and pier foundation.

For shallow foundations，isolated footing refers to the single footing or column footing which is used in framed structures where several columns are there to be supported. Combined footing，the other type of shallow foundation，is formed by joining two neighboring columns since little or no space exists between them so that the two footings have to overlap with each other.

When it comes to deep foundations，the pile foundation is one important type and commonly constructed one on sites，particularly for deep excavation sites. Whatever materials piles are made of，they are installed deep into the ground. Some piles are pre-casted in a factory and driven into the ground by heavy machines. Others are cast-in-situ by drilling holes and filling them with concrete. When a group of piles are driven into the soil，the top of the piles must be at the same level and capped with concrete slab or other pile caps on which superstructures are built. In this way，loads from superstructures can be transferred to the soil through piles either by friction between piles and soil or by direct being beard by hard strata. That is why there are load-bearing piles and friction piles. Since piles are driven into the ground，it is inevitable for them to be damaged during this process， apart from being damaged during handling and transportation if they are precast piles. To check if they are damaged and by how much, tests must be done according to the requirement of the code which include dynamic load test, static load test, and pile integrity test.

The other type of deep foundation is piers, they are a group of large diameter cylindrical columns which support the superstructure and transfer their loads into the hard strata deep in the ground. They are above ground, several meters high.

(3) Words and Phrases

substructure 基础，根基，下部结构，下层建筑
superstructure 上层建筑，上层结构
overturn 倾覆
shallow foundation 浅基础
isolated footing 独立底脚，独立基础
combined footing 联合底座，联合基础，复合柱基
deep foundation 深基础
pile foundation 桩基础
pier foundation 墩式基础
single footing 单独底脚，独立基础
column footing 柱脚
frame structure 框架结构
pre-casted 预制的
cast-in-site 现场浇筑
pile cap 桩帽
strata 岩层
friction 摩擦，摩擦力
load bearing pile 承重桩
frictional pile 摩擦桩
dynamic load test 动载荷试验
static load test 静载荷试验
pile integrity test 桩完整性试验

(1) 导读

众所周知，基础是建筑物中极其重要的一个组成部分。基础施工的质量关系到整个建

筑物的总体质量水平。本篇将着重介绍基础的定义、功能和类型。

（2）正文

基础是建筑物中处于底部的那部分结构组件，也称为下部结构。基础施工是一个过程，它是指在地面以下进行开挖，使其达到设计所要求的基坑形状和标高，然后浇筑出一个大的结构体。它是施工项目中一个重要的组成部分，同时也是第一个施工活动，随后其他施工活动可以在其基础之上相继展开。基础的功能一是把来自上层结构的荷载进行分散并传输到地下的土壤，二是抵抗作用在建筑物上的各种荷载（包括建筑物自身的横向荷载和风荷载），防止这些荷载使建筑物产生移位和倾覆。此外，基础也提供了一个水平的作业工作面，所有其他的结构可以在其上面进行施工并得到它的相应支撑。

基础有两种类型：浅基础和深基础。浅基础又可进一步细分为独立基础和联合基础，而深基础可细分为桩基础和墩式基础。

浅基础中的独立基础是指单独基础或柱基础，它通常用在框架结构型的建筑物中，在这些建筑物中存在一些需要被支撑的柱子。另外一种类型的浅基础为联合基础，它是将两个相邻的柱子浇筑成一个整体，如此浇筑施工是因为两个柱子之间几乎没有空间，所以两个基础也就自然混为一体。

深基础中的桩基础是一个重要的类型，它一般在现场浇筑而成，尤其是在那些深基坑现场。无论桩的组成材料如何，它们都被安装在地层的深处。部分桩是在厂家预制好之后由压桩机直接压入土中的，而其他的桩则是在施工现场通过钻孔和浇筑混凝土而形成的。当一组桩被压入土壤深层时，所有桩的顶部应该保持在同一标高上，并且应在这些顶部上面安装相应的混凝土板或其他类型的桩帽，这样就可以在这个新的平面上进行上层结构的作业。通过上述施工工艺，来自上层结构的荷载就可以通过桩传递给土壤，荷载可以通过桩和土壤之间的摩擦来进行传递，也可以由硬岩层直接来承受，因此也就有摩擦桩和端承桩这两种类型的桩。在压桩过程中，有些桩可能会受到损坏，预制桩还极有可能在运输和装卸过程中受损。为了检查桩有无受损和判断受损的程度大小，应该按照规范要求进行一些试验，包括动载荷试验、静载荷试验和桩基完整性试验。

另外一种深基础就是墩式基础，它们是一组大直径的圆柱，用来支撑上层结构并将上层结构给它们的荷载传递至位于地表深层的硬岩层。墩位于地面之上，一般有几米高。

2. Reading：Footing Near Slope（边坡基础）

（1）Introduction

Apart from frostproofing of footing，the nature of the soil surrounding and under it will play a great role when it concerns its stability. What precautions should be taken as well as how to do them when building a footing near a slope will be discussed as follows.

(2) Text

There is no doubt that all buildings are built upon and supported by a foundation or footing, footing is one type of foundation. Footing can be constructed in different kinds of terrain, such as a seaside, open county, mountain areas, and hillsides. So, when it is needed to build a house near a hill, what measures should be taken to protect its footing? What is frost precaution during and after concreting the footing?

When building a footing on or near a slope, enough clearance from the ascending slope or setback from the descending slope should be allowed between the footing and slope. Sometimes, a retaining wall is required to protect footing or structure from slope drainage, erosion, and other damage caused by the slope. Due to the nature of the soil surrounding the footing, its top surface must be leveled well whereas the bottom can have a slope. If it is necessary to change the elevation of the top surface of a footing or some special ground slopes are encountered, footing could have several steps. When it comes to depth of footing, 12 inches should be the minimum requirement below undistorted ground. In particular, build a footing with sufficient depth to ensure enough stability if soil with a character of shifting or moving is encountered. Footing should be constructed upon the soil, and the maximum bearing capacity of soil should not be exceeded. That is all about the shape, and depth of footing. Let us move on to frostproofing.

Indeed, frostproofing is a measure to be taken to protect a footing from freezing during concrete placing and for at least five days thereafter. Most importantly, no running water is allowed on the surface of freshly concreting footing. Of course, formwork is not needed if the soil condition is firm enough for the concrete to be cast against. On the whole, frost precautions should be taken for foundation walls, piers, and other permanent supports of structures.

Lastly, as for a steel grillage footing, approved steel spacers should be installed to separate the structural steel shapes and the whole grillage footing should be encased totally in the concrete with proper coverage on all sides. Concrete or cement grout should fill in the spaces between the shapes.

(3) Words and Phrases

terrain　地形

top surface　顶面

undistorted ground　未经挖动的土地（地面）

frost precaution　防冻措施

slope 斜坡，坡道

setback 收进，后退的最小距离（房屋从地界后退的）

retaining wall 挡土墙，拥壁

bearing capacity 承载能力

steel grillage footing 钢格排基础

spacer 水泥垫，定位架，调整垫

（1）导读

除了基础的防冻情况之外，它的周围及其下面土壤的土质也极大地关系到基础的稳定性。本篇将着重介绍在斜坡附近浇筑基础时应该采取什么样的预防措施以及如何实施这些预防措施。

（2）正文

毫无疑问，所有的建筑物都是建造在基础之上的，并且由基础来提供支撑。基础的建造环境可以有各种不同的地理特征，例如海边、平原、山区和山坡边等。因此，如果必须在山的附近来建造房子的话，应该采取什么样的措施来保护基础？在浇筑基础的过程中和之后的一段时间里，如何对基础进行防冻保护呢？

在山坡上或其附近建造基础时，应该使基础和山坡的坡身之间的距离大于一定的最小距离，这个最小距离应该适应上行坡和下行坡。特殊情况下，需要建造一个挡土墙来保护基础或建筑物，以免其受到损坏，可能的损坏包括山坡排水、山坡冲刷侵蚀以及其他源于山坡地理特征的损坏。由于基础附近山坡的自然地理特征的局限性，它的底部可以为坡形，但它的顶面必须进行找平。假如基础顶面的标高确实有必要改变或者是遇到了一些特殊的地表情况，基础的顶面可以有不同标高的台阶。至于基础的深度，最低要求应是达到未经挖动的地面以下 12 英寸。特别是当土壤有移动或滑动的潜在危险时，建造的基础必须要有足够的深度来确保基础的稳定可靠。显然，基础应该建造在土壤之上，且不应该超过土壤的最大承载能力。上面讲到的是基础的形状和深度，下面再来谈一下基础的防冻问题。

事实上，防冻是在基础浇筑期间和随后至少五天的时间里需要采取的一些措施和方法，以保护基础免于冻坏。更为重要的是，在刚刚浇筑完的混凝土的基础表面上不允许有任何形式的积水存在。当然，如果土壤的土质很坚固，使得混凝土可以直接在其上进行浇筑的话，那么也就没有必要安装模板了。总之，基础墙、墩式基础和建筑物其他类型的永

久支撑都应该采取防冻措施来对它们进行保护。

最后，对于钢格排基础，应该安装经过批准的钢质调整垫来将钢结构组件分隔开来，整个钢格排基础应该全部浇筑在混凝土里，并且要保证它所有的外表面都具有设计所要求的混凝土保护层，钢结构组件之间的缝隙应该填充混凝土或水泥灌浆。

3. Reading：Dewatering （降水）

(1) Introduction

It is not uncommon to do dewatering in most construction sites, deep foundations in particular. it is crucial to have a dry and sound working condition for the foundation so the reasons and types of dewatering will be discussed here.

(2) Text

Types of structures, such as powerhouse, dam, multistory building, and basement, can be different, but one thing in common among them is that all these structures are supported by and built upon the foundation or footing. To construct a foundation, excavation, either shallow or deep, must be executed. Most of the time, excavation might be done into water-bearing soil, so there is a need to lower the water table to avoid seepage of water. Besides, some running water on the working surface at the bottom of the excavation is not good for the safety of workers and the quality of structures to be built, they also need to be channeled away properly. Therefore, dewatering is essential regarding having a dry and safe condition for excavation.

Dewatering is the process of removing surface and subsurface water from a construction site. More often, it consists of a collection of water, pumping, filtering/removing silt, and discharge. If dewatering is done properly, it will do an easy and safe accomplishment of work very well.

By removing a small amount of water in case of shallow excavations up to 1.5m deep and controlling a large amount of water in deep excavations over 3m deep, dewatering can prevent leakage of water or sand as well as disturbance of the soil at the bottom of the excavation, therefore increasing the stability of soil and having good water stability in the girth. By doing this, it has a good chance to have a dry base for the foundation, as well as a safe and sound working condition for workers and machines. Dewatering is so important for construction work that an appropriate method of dewatering must be chosen for a specific project, due to the nature of the soil, groundwater condition, and risk assessment.

Generally speaking, there are four methods of dewatering: open sump pumping, well point system, deep well system, and eductor system. Here open sump pumping and well point system will be discussed.

First，open sump pumping has an area in the ground deeper than the basement where water is collected and pumped away for disposal. It is the most reliable，simplest，cheapest，and most effective way of dewatering. It consists of collector drains，sump，sump pump，and discharge pipe（as in Fig. 3-1）. Collector drains and sumps are built at one or more sides or corners of the foundation pit，the drains collect the groundwater and convey it into the sump from which the water is continuously pumped out by the sump pump through a discharge pipe. Second，well point system consists of a series of small-diameter well points，header pipe，discharge pipe，settlement tank，and dewatering pump. Well points are connected to a header pipe which leads to a pump. The groundwater is drawn by the pump into the well points through the header pipe before being finally discharged into the settlement tank. This type of dewatering system is suitable and effective in sands and sandy gravels. In a word，choose the right method of dewatering for given construction work，otherwise wrong method of dewatering will cause surface flooding which would damage adjacent properties and erosion，and other related problems.

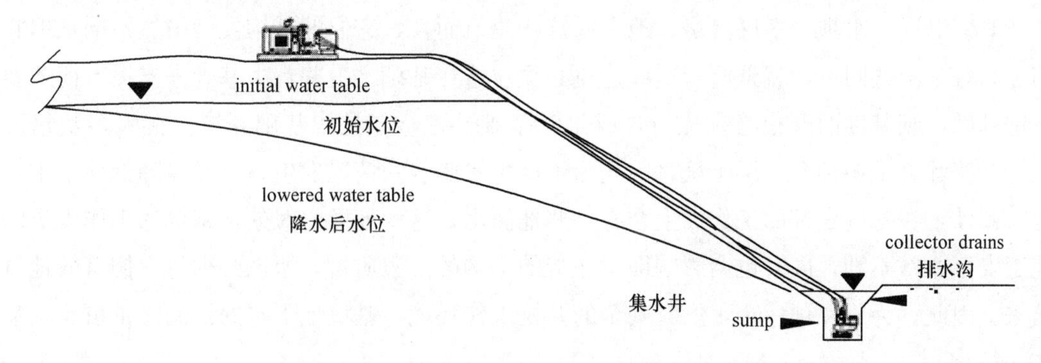

Fig. 3-1 Open Sump Pumping 基坑明沟排水

（3）Words and Phrases

multistory building 多层建筑

water table 地下水位

seepage 渗水

dewatering 降水

subsurface 地下

removing silt 清除泥沙

girth 周围（尺寸，长度），场合

open sump pumping 基坑明沟排水

well point system 井点降水

deep well system 深井井点降水

eductor system　喷射井点降水
collector drain　排水沟
sump　集水井

（1）导读

在大部分的施工现场，基坑降水是一种非常普遍的施工活动，尤其是在一些深基础的施工环境。一个适宜的工作环境和不受地下水的干扰与影响，对基础的施工至关重要。本篇将着重介绍基坑降水的原因和一些方式方法。

（2）正文

对于电厂、水坝、多层楼房、地下室这些建筑而言，它们的结构类型虽然不同，但它们之间有一个共同点，就是所有这些建筑必须建造在基础之上并且由基础来支撑。而若要建造基础，则基础的开挖必须先行，或是深基础开挖，或是浅基础开挖。在大多数情况下，开挖都会挖掘到含水层土壤，因此为了避免渗漏水的情况发生，有必要降低地下水水位。此外，在基坑底部的工作面上会有一些地面水，这些地面水对处在基坑的工作人员的施工安全非常不利，同时也会影响即将建造的结构的工程质量，所以必须将它们有效地排出去。因此，为了保证一个干燥且安全的开挖工作环境，基坑地下水的控制必不可少、势在必行。

地下水的水位控制实际上就是将地面水和地下水从基坑的施工现场排出坑外。在大部分情况下，它包含收集水、水泵抽水、过滤和清除砂土、排放等过程。水位控制得好将非常有利于工程的安全施工和顺利进行。

在进行深度不足 1.5m 的浅基础开挖时，排水量较少；而在进行深度为 3m 及以上的深基础开挖时，则需要控制大量的地下水，通过地下水的水位控制可以有效地防止地下水和砂土的渗入以及基坑底的隆起，从而进一步增强土壤和周围环境中地下水的稳定性。如果能这样做的话，就很有可能保证基础浇筑在一个干燥的工作面上进行施工，施工工人和设备都能有一个安全且可靠的工作环境。鉴于地下水位的控制对于施工活动的重要性，以及任何一个项目现场的土质情况、透水层位置和存在的风险情况各不相同，选择适于其项目自身特点的降水方法就显得尤为必要。

一般来讲，有四种基坑降水的排水方法：基坑明沟排水、井点降水、深井井点降水和喷射井点降水。这里鉴于篇幅有限，只简单介绍基坑明沟排水和井点降水。

基坑明沟排水是指有一个比基坑地表面低一些的集水井，在这里水被收集起来并由泵排出，之后再进行相应的处理，这种降水方式最可靠、简单、经济和有效。其装置由排水沟、集水井、泵和排水管组成（图 3-1）。排水沟和集水井设置在基坑的周边或角落里，基坑里的水沿排水沟流到集水井里，然后再由排水泵不停地抽出，最终经排水管排放到坑外。另一种排水方法为井点降水，它的装置由一系列直径很小的井点管、集水总管、排水管、泥浆池和水泵组成。井点管和集水总管相连接，而集水总管又与水泵连接。地下水由水泵抽到井点管后再经由集水总管被排放到泥浆池里，这种降水方式更适用于砂土地基。总之，针对特定的施工项目应采取相应适宜的降水方法，错误的降水方法会导致基坑的作业面积水，进而对邻近的建筑造成损坏、侵蚀或导致其他问题。

4. Dialogue：Boring Pile Wall（钻孔灌注桩）

At 3 p. m. on Tuesday, on one excavation site, Jack is showing David around the site of a boring pile wall. David is an intern.

David：I see some wooden pegs over there, what are those pegs for?

Jack：Oh, those wooden pegs are driven down there and used to mark out the center position of each bored pile.

David：By whom?

Jack：By our surveyors, they do those things before the installation of the casing.

David：Casing?

Jack：Yes, look over there, the Vibro Hammer is driving a casing into the ground.

David：How deep downward will the casing go?

Jack：Till 1-meter length of the casing protrudes from the ground.

David：What happens after that?

Jack：Let us walk to the Auger and have a look.

David：Too much of earth here.

Jack：You are right, you see, the auger, a drill tool, is cutting and removing the soil within the casing to form a borehole.

David：Why is a borehole needed?

Jack：Because it provides a room, a crane will lift the steel cage and place it within that room.

David：Can the pump truck pour concrete into the borehole to form a bored pile after that?

Jack：Yeah, that is what the pump truck over there is doing now.

David：Once the bored pile is formed, shall the casing still be left there?

Jack：No, Vibro Hammer will extract the casing out from the ground for another bored pile.

David：Does a new cycle begin?

Jack：Yes，it does，till the entire length of the CBP (Contiguous Bored Pile) wall is completed.

David：It is 3：30 already. Do they keep doing this work throughout the night?

Jack：No，they stop working at 5 p. m. They time their work carefully and try their best not to make noise and vibration at night，disturbing the life of people around.

David：Are the noise and vibration monitored?

Jack：Of course，some instruments are installed in the vicinity of the excavation site to monitor vibrations and ground movements，etc.

David：I see.

Jack：Let us go there，I show you some instruments.

周二下午三点，在一个基坑的开挖现场中，杰克正在带领实习生戴维参观钻孔桩的施工现场。

戴维：我看到那边有一些控制桩，它们是做什么用的？

杰克：啊，打那些控制桩是用来标识每个钻孔桩的中心位置的。

戴维：是谁打的控制桩？

杰克：是我们的测量员。在安装套管之前，测量员在测量之后打控制桩。

戴维：套管？

杰克：对的。看那边，振动锤正在把一个套管往地下打。

戴维：套管要打到多深？

杰克：一直要振打到套管的顶部露出地面 1m 时。

戴维：接着下面是什么施工步骤？

杰克：我们到螺旋钻机那边去看一下。

戴维：这里怎么有这么多的土方呢？

杰克：是的，钻机在套管里把土方粉碎清除出来之后就形成一个个钻孔。

戴维：钻孔的用途是什么？

杰克：钻孔实际上是一个地下空间，吊机把钢骨架吊起之后把它放进钻孔里。

戴维：然后泵车把混凝土浇灌进钻孔里，形成一个钻孔桩？

杰克：对的。混凝土泵车正在那边浇灌呢！

戴维：钻孔桩浇灌好后，套管还留在地下吗？

杰克：当然不是，振动锤会把套管从地下提取上来，以便打下一个桩时再使用。

戴维：然后再开始下一轮的施工程序？

杰克：是的，再开始浇灌下一个钻孔桩。一直循环下去，直到所有钻孔桩浇筑完毕。

戴维：已经三点半了，他们晚上还加班吗？

杰克：不，他们五点钟停工。他们在安排施工的进度时非常谨慎，尽量避免夜间施工产生噪声和振动，干扰周边居民的生活。

戴维：噪声和振动受到监控吗?

杰克：是的，在开挖现场附近安装了一些监测探头，它们会随时监控施工现场产生的噪声和振动这些环境方面的问题。

戴维：我明白了。

杰克：我们到那边去，那里安装了一些监控探头，你可以看一下。

5. Exercises （练习）

(1) Some large and small diameter bored cast-in-place __1__ are often used to construct efficient and economic temporary or permanent retaining walls.

(2) The choice of bored pile wall systems depends on many factors, these include soil __2__, groundwater profile, retained heights, available construction time, propping arrangement, cost, and design __3__.

(3) Contiguous piles are suitable where the groundwater __4__ is below excavation level. it is normally the most economic and rapid option.

Unit 4

Formwork （模板系统）

1. Reading：Formwork and Its Materials（模板系统和材料）

(1) Introduction

On today's construction site, it is almost impossible to build a concrete structure without formwork. Formwork materials cost a lot when it comes to construction costs on site, here the definition of formwork and its materials will be discussed.

(2) Text

Formwork is a system of connected forms, consisting of sheeting, bearer, bracing, ties, and other accessories. On a construction site, it is designed and constructed to support structures and molds to create concrete structures of desired size and shape. Therefore, on one hand, it should be strong enough to support dead and live loads imposed on them. On the other hand, it should be easy to be removed when the concrete structures are completed. Since it needs lots of time and money to construct and then remove, causing major costs and safety concerns among concrete works, it is the responsibility of both engineers and construction staff to design, erect, strip, and maintain the formwork economically and safely.

Formwork comprises form-facing materials, form accessories, formwork release agents, expansion joint fillers, and other embedded items. First, form-facing materials could be plywood, tempered concrete-form-grade hardboard, metal, plastic, paper, or other appropriate materials depending on the purpose of the formwork. They should produce a good finished concrete surface of smooth, uniform texture without blemish. Second, formwork accessories might include ties, hangers, clamps, yokes, nuts, and others. They are used to fasten and secure the formwork. Third, formwork release agents are used to prevent forms from absorbing moistures from fresh concrete, avoid bonding between concrete and sheeting of formwork, and minimize the concrete stains on the concrete surface. They are classified into water-based and solvent-based. Next, expansion joint fillers could be asphalt, fiber expansion joint, and others. They are used to bind and hold parts together in concrete, absorb the vibration, and even allow the movement of parts during an earthquake. Finally, other embedded elements could be either rubber water-stop or polyvinyl chloride water-stop. Water-stop is used to provide watertight for concrete works.

In short, formwork consists of different kinds of materials to form a mold in which fresh concrete is poured, gains its strength, and finally forms the required structure of desired size and shape.

(3) Words and Phrases

sheeting 挡板，板栅
bearer 支撑，垫块，承木
bracing 支撑
tie 拉杆，横拉撑
dead load 恒载，静载，静重
live load 活荷载
facing material 饰面料，覆面材料
form accessory 模板附件，备件
formwork release agent 隔离剂
expansion joint filler 伸缩缝填料
hanger 吊钩，支架
clamp 夹，夹紧装置
yoke 叉臂，刀杆支架
embedded element 预埋构件
water-stop 止水带

译文

(1) 导读

如今的施工现场，如果没有模板系统，人们几乎无法建造任何形式的混凝土结构。施工成本中，模板材料费用通常占据很大比例。本篇将着重介绍模板系统的定义和组成材料。

(2) 正文

模板系统是由模板连接起来的，其组成部分有挡板、垫块、支撑、横拉撑和其他附件。在施工现场，经过周密设计和安装的模板支撑混凝土结构并形成模型，在这个模型里面浇筑出位置尺寸和几何形状满足设计要求的混凝土结构或构件。因此，一方面，模板自身的强度应该能够支撑承受作用在它们之上的静荷载和动荷载；另一方面，在混凝土结构浇筑凝固后，它们应该能够被轻易地拆除掉。由于模板的安装和拆除需要花费大量的时间和资金，模板工程造价是钢筋混凝土工程施工费用中最大的一笔支出，同时模板施工也是最容易出现安全隐患的环节，设计人员和工程施工人员都有责任和义务从更经济、安全的

角度来进行设计、安装、拆除和维护模板。

模板系统由模板、附件、隔离剂、伸缩缝填料和其他预埋构件组成。第一，根据模板系统的用途，模板可以是胶合板、混凝土模板用的高强度耐水硬纤维板、金属、塑料、纸或其他适宜材料。浇筑后的混凝土结构的表面应平滑、纹理一致且无缺陷。第二，模板附件一般有拉杆、钩头螺栓、柱箍、对拉螺栓、螺母及其他备件，它们用来将模板固定牢固。第三，隔离剂可以用来防止模板的挡板吸收刚浇筑好的混凝土里的水分，预防混凝土粘结在模板表面，减少混凝土表面出现的污点，隔离剂有水基和油基两种。第四，伸缩缝填料可以是沥青、纤维伸缩接头以及其他材料。它们用来将建筑结构的各个构件结合起来使之成为一个整体，能够减少振动，甚至在地震时可以允许建筑结构的构件产生一定的位移。最后，其他预埋构件可能是橡胶止水带或者聚氯乙烯止水带，止水带能够保证混凝土的水密性。

总之，由各种材料组成的模板可以作为一个模型，在其中浇筑新拌混凝土，直至混凝土凝固达到规定的强度要求，最终形成符合设计所要求的位置尺寸和几何形状的混凝土结构体。

2. Reading：Types of Formwork（模板系统类型）

（1）Introduction

Formwork can be classified into different groups according to its usefulness and physical position. It consists of several elements with different names due to its various applications. Here types of formwork will be discussed.

（2）Text

Formwork can be further broken down into panel forms and stationary forms according to its usefulness. Panel forms are reusable while stationary forms are not reusable.

Another way of categorizing formwork is based on its physical location：horizontal forms and vertical forms. Horizontal forms are used to build bridge decks，floors，overhangs，balconies，and other horizontal elements. Also，they should be capable of withstanding the weight of concrete，formworks themselves，workers，and equipment. On the contrary，Vertical forms are employed to produce walls，columns，piers，abutments，and other vertical concrete surfaces，which shall be good enough for holding the pressure from concrete，wind force，forces caused by vibration of a vibrator，and others.

Firstly，formwork for walls consists of timber sheeting，vertical posts，horizontal members，rackers，stakes，and wedges. Secondly，formwork for columns contains side and end planks, yokes, nuts, and bolts. Two end and side planks are joined by the yokes and bolts. Thirdly, formwork for slabs and beams comprises sole plates, wedges, props, head tree, planks, battens, and ledgers. Beam formwork rests on a head tree while slab formwork on battens and joists. Fourthly, and last, formwork for stairs is made up of vertical & inclined posts, inclined members, wooden planks or sheeting, riser planks, and stringers. No matter what they are, horizontal or vertical, forms should be constructed and dismantled under the supervision of a competent person according to the building plan and code requirements.

In general, materials used to construct formworks include timber, metal, plastic, and others. To start with, timbers are easy to process, like cutting and sawing. Timber formworks are still popular on today's construction sites since plywood and hardboard are light to handle and simple to be replaced, as well as have good insulation properties. If labor cost is not a major concern compared with purchase cost, timber formwork is the best option. Secondly and equally importantly, metal formworks include steel formworks and aluminum formworks. Unlike timber formworks, both steel formworks and aluminum formworks are strong and durable and have a greater number of times for reuse as well as no shrinkage. The difference between steel formworks and aluminum formworks is that aluminum formworks cost much more than steel formworks due to their high speed of construction. Finally, plastic formworks are light in weight, reusable, and chemical resistant, while drawbacks are that they are more expensive for the first time and could not withstand high loads.

When it comes to choosing which kind of materials should be utilized for formwork, the costs of materials, nature of the project, schedule of construction, safety, and other issues should be taken into consideration.

(3) Words and Phrases

panel form 格板模板
stationary form 固定模板
deck 桥面，平台
overhang 突出，外伸，悬垂
abutment 桥台，桥墩
racker 斜向支撑
stake 木桩
wedge 楔块
riser plank 踏步竖板
stringer 楼梯斜梁

(1) 导读

模板根据其用途和使用场合可分为不同的类型。模板是由几个不同的组件组合而成的，各个不同的组件有着不同的用处。本篇将着重介绍一下模板的类型。

(2) 正文

根据模板能否被重复使用这一特点，它可以进一步划分为格板模板和固定模板两种类型，其中格板模板可以被拆移并重复使用，而固定模板不能被重复使用。

模板的另一种分类方式是按照其具体使用场合的特点进行分类，分为水平模板和竖向模板。其中水平模板通常用于建造大桥的桥台、地板、外伸平台、阳台和其他水平构件，它们应能支撑混凝土、模板自身、工人和设备等所有的荷载。而竖向模板通常用于墙、柱、墩、桥台和其他垂直混凝土面的浇筑，它应能承载来自混凝土、风力、振捣器产生的振动荷载和其他的横向荷载。

第一，墙模板系统包括木挡板、立管、水平杆、斜向支撑、木桩和楔块等。第二，柱模板系统由端模板、侧模板、对拉螺栓、对拉螺杆和螺母等组成，其中端模板和侧模板由对拉螺杆紧固起来。第三，楼板模板和梁模板这两种模板系统由垫木、楔块、支撑、帽木、底模板、侧模板、木方和托木等组成。梁模板安装在帽木上，而楼板模板安装在木方和连接件上。第四，楼梯模板系统包括支撑立杆、斜撑、反三角板、楼梯底板、踏板和斜木楞等。此外，无论是水平模板还是竖向模板，都应由具有执业资格的工程人员依据施工图纸和规范要求来监督模板的安装和拆除。

一般来讲，用于安装模板系统的材料有木板、金属、塑料和其他适宜的材料。首先，木板容易被加工，如进行锯切处理。木模板系统在施工现场仍然被普遍使用，其主要原因是胶合板和硬质纤维板这两种材料质量轻、操作容易、替换简单且保温效果好。如果与材料成本相比，人力成本不是主要考量因素的话，木模板系统应是最佳选择。其次，类似地，金属模板系统有钢模板和铝模板两种模板系统。与木模板系统不同的是，钢模板和铝模板这两种模板系统都很坚固耐用，可以被多次周转使用，且不会因水收缩而变形。钢模板系统和铝模板系统之间的区别是，铝模板系统比钢模板系统的成本要大很多，其施工进度更快。最后，塑料模板系统质量轻、可被重复使用且耐化学腐蚀，其不足之处是首次使用的费用较高一些，同时承载重负荷的能力较弱。

至于选择哪种材料作为模板材料，应充分考虑以下一些因素：材料成本、项目特点、施工进度、安全和其他相关问题。

3. Reading：Construction of Formwork （模板系统搭建）

（1） Introduction

Formwork costs a great amount of money and has something to do with the quality of a concrete structure. It is crucial to erect it correctly according to the best practice. Here the typical procedures for preparing and erecting will be discussed.

（2） Text

Usually，the construction of formwork contains propping and centering，shuttering，provision of camber，cleaning，and surface treatment. Take a column as an example，its erection sequence could be as follows：

1） Prepare

① Check the steel formworks for their defects on the surfaces and cleanness.

② Clear working area and surfaces of formwork before the erection of formwork.

2） Set out

Locate and place the formwork on the required grids according to the building plan.

3） Assemble

① Put side and end sheeting board of formwork together based on the plan with supports set to the right height level and line.

② Brace the formwork board and secure them tightly.

③ Install block-outs and cast-in services in line with the building plan.

4） Check

① Plumb forms both ways and fasten them by using adjustable props.

② Examine and make sure that the propping angle is 45 degrees to the floor.

5） Clear within the Formwork

Clear all debris，sawdust，and other foreign materials within and around formwork.

6） Apply Release Agents

Apply formwork release agents to the formwork surface as per instruction from the manufacturer.

7） Final Inspection before Concreting

Inspect the plumbness and alignment of formwork before concreting.

（3） Words and Phrases

propping 支撑

centering 拱模

surface treatment 表面处理

erection 竖立，架设

block-out 暗盒埋件

plumb bob 铅锤，测锤

译文

（1）导读

模板系统的成本较高，且容易影响浇筑的混凝土结构的质量，因此，按照同行业最佳的施工工艺正确安装模板就显得至关重要。本篇将着重介绍安装模板系统的前期准备和具体施工过程。

（2）正文

一般来讲，模板系统的安装包括支撑及水平模板的安装、竖向模板的安装、梁拱的搭设、场地的清洁和模板表面的处理。以柱子为例，其安装步骤如下：

1）准备工作

① 检查钢模板表面有无缺陷及清洁程度。

② 为模板系统的安装清理、打扫出工作区域和模板表面。

2）放线

依据施工图纸进行测量，并将模板放置在要求的位置线上。

3）安装

① 依据施工图纸将柱子的各块模板依次安装就位并做好临时支撑，保证高度和位置线符合设计要求。

② 用柱箍将木模板固定牢靠。

③ 依据施工图纸在适当的位置和高度安装预埋预留孔堵块和管线。

4）检查

① 用线锤来控制和校正模板的垂直度，并用可调节的拉杆或斜撑来固定模板。

② 检查并确保斜撑和地面之间的角度为 45°。

5）模板内清扫

在模板内部及其外围清扫所有的杂物、木屑及其他建筑垃圾。

6）喷刷隔离剂

按照厂家的产品说明书要求在模板表面喷刷隔离剂。

7）混凝土浇筑前的最终检查

在浇筑混凝土之前检查模板的垂直度及模板与柱控制线的偏差。

4. Dialogue：Scaffolding Erection （脚手架搭建）

At 10 a. m. on Tuesday，on one construction site，Jack and David are at the scaffolding site，together with scaffolders.

Jack：They are now erecting a scaffold for an office building.

David：I see，the area is barricaded，there is warning signage for scaffolding.

Jack：The whole working area is clearly defined.

David：Is this a common practice on scaffolding sites?

Jack：Indeed，yes，it is. Scaffolding is a dangerous job. Signage shows that scaffolding is being erected and is not yet safe for use.

David：A lot of components there，have they already received all the correct components?

Jack：I guess so，they must have counted and checked off the various parts against a delivery docket.

David：How is scaffolding erected?

Jack：It takes time to do it. First，roughly set up the positions of the sole plates. By the way，the sole plates are leveled.

David：What is next?

Jack：Place ledges and transoms between two sole plates.

David：Anything special to take care of here?

Jack：Use the base plate to compact down the sole plates and make sure the sole plates form a firm foundation.

David：What if the ground is uneven?

Jack：Adjust the base plates to compensate for it.

David：Now the first bay can be erected?

Jack：Yeah，insert the base plates into the standards，do this on both sides.

David：Scaffolding is teamwork.

Jack：Right. You see，the standards are lifted，and one transom is inserted into the lowest V connectors.

David：Why has the standard on the right been rotated?

Jack：Because by doing so，the highest V connectors can go along the faces of the scaffold.

David：They are holding the standards at the other end of the bay in place and attaching the ledges.

Jack：Yes，they will insert the final transom to link the standards.

David：Will this bay be leveled?

Jack：They will do it by using a spirit level.

David：How to check that the bay is square?

Jack：An easy way is to see how the planks align with the transoms.

David：Is this the last step for the first bay?

Jack：Almost done，the pins and wedges are hammered into place，then the upper transom and ledges are installed and fixed off. The first bay is completed.

David：Will they carry on the rest of the scaffold after that?

Jack：Yes，they will.

对话

周二上午10点，杰克和戴维在一个脚手架施工现场检查工作，架子工们正在搭设脚手架。

杰克：他们正在为一个办公楼的施工现场搭设脚手架。

戴维：是的，施工区域已经被隔开了，还悬挂着“脚手架正在施工”的警示牌。

杰克：这样做的话，整个工作区域就能被清晰地划分出来了。

戴维：这些措施都是脚手架施工现场的一些常规做法吗？

杰克：是的，确实是这样。脚手架的搭设是一项危险工作，警示牌可以提醒人们脚手架正在搭设，现在还不能安全地使用。

戴维：那里有很多零备件，他们已经收到所有需要用的零备件了吧？

杰克：应该是的，他们一定已经按照发货清单来清点和核对过各个零备件了。

戴维：脚手架是如何搭设的？

杰克：搭设脚手架需要一定的时间。首先，把垫板放置在大致的位置上，当然，垫板下面的地面应该进行找平和夯实。

戴维：下面接着做什么呢？

杰克：把纵向水平杆和横向水平杆放置在两个垫板之间。

戴维：这里有什么需要特别注意的事项吗？

杰克：有的，注意要用底座把垫板向下夯实，确保垫板能够形成一个坚实的基础。

戴维：如果地面不是在同一地坪上呢？

杰克：那么就要调整底座来补偿地面的高低差。

戴维：现在可以安装第一个架体了吗？

杰克：是的，要把底座插入立杆里，两边采用同样的操作方法。

戴维：我觉得搭建脚手架需要一个团队来协同合作。

杰克：是的。你瞧，他们正提起立杆，把横向水平杆插入最下面的V形接头里。

戴维：右边的立杆为什么要转一下呢？

杰克：因为这样做之后最上面的V形接头就可以和脚手架的立面保持一致了。

戴维：他们正在把架体另一端的立杆放到安装位置上，并用一根纵向水平杆把两根立杆连接起来。

杰克：是的，接着他们将插入最后一根横向水平杆来把立杆连接在一起。

戴维：这里的第一个架体还需要测量水平吗？

杰克：需要的，他们会使用水平仪来测量架体的水平。

戴维：如何检查架体是否歪斜呢？

杰克：一个简易的方法是检查脚手板是否和横向水平杆都对齐了。

戴维：这是第一个架体的最后一项操作了吗？

杰克：是的，它几乎就要完成了。再用锤子把销子和楔子锤进插孔里，然后把上层的纵向水平杆和横向水平杆安装入位，这样第一个架体就算完工了。

戴维：接着他们会搭建第二个、第三个等其他的架体吗？

杰克：没错，是这样的。

5. Exercises（练习）

(1) Scaffold inspections must be completed every 30 days for scaffolds with a __1__ risk of more than 4 meters.

(2) If an inspection identifies a problem with the scaffold, __2__ to the scaffold must be controlled and any necessary repairs, alterations, and additions completed.

(3) Once a scaffold has been erected a handover inspection should be completed to check that the scaffold is __3__ for use.

(4) __4__ access to the scaffold is prevented while the scaffold is incomplete or unattended.

(5) The scaffold and its supporting __5__ must be inspected by a competent person before the scaffold is used after an incident has occurred that might affect the stability of the scaffold.

Unit 5

Rebar （钢筋）

1. Reading：Rebar Test （钢筋试验）

(1) Introduction

Rebars are used a lot on construction sites and affect the quality of constructed concrete structures，therefore their test concerns engineers very much. Here samples and test of rebars are discussed as follows.

(2) Text

Rebars are produced around the world. The question is how to decide whether the rebars produced by a particular mill would meet the requirement of codes and specifications. The answer is established by taking samples of test specimens of rebars and testing them in the lab.

The number of test specimens per batch should be decided based on the requirement of local codes and by-laws or instruction from engineers. Once the number of samples is calculated，the required amount of test specimens should be cut from the given batch on-site and taken to the laboratory for test.

Tests normally consist of tensile test，bend test，and fatigue test. The standards and codes as well as local by-laws would say how to do these tests and how to interpret the test results. The types of tests needed for rebars depend on where the project is located and what codes the project must follow. Usually，the following information could be found from those codes and specifications：equipment to be used，the definition of some words and symbols，how to prepare the samples，methods of test，and results.

After the test report is received，the report should be interpreted by a competent person to decide what to do next. If the report results do satisfy the requirement of codes，it is appropriate to proceed with the next work. If the results of the report do not，then take further samples from the same patch and repeat test. As for how many samples should be taken at this time，local codes should be referred to and engineers consulted.

Rebars tests should be done in authorized laboratories designated by engineers and representatives of the owner with the test reports submitted to them for decision making. Test reports must be kept for records.

(3) Words and Phrases

mill 厂家

test specimen 试验样件

batch　批次
tensile test　拉伸试验
bend test　弯曲试验
fatigue test　疲劳试验

（1）导读

钢筋在施工现场的应用十分广泛，并且影响建造混凝土构件和结构的质量，因此现场的工程人员对钢筋的检测极为重视。本篇将着重介绍钢筋检测的取样方式和种类。

（2）正文

生产钢筋的厂家遍布世界各地，为了确保某一厂家生产的钢筋满足工程所指定的规范和标准的要求，需要把它们的钢筋按生产批次进行取样，并送到试验室进行检测。

每批钢筋的抽样方案应依据项目当地的相关法律法规或者所咨询的工程师的意见来决定。一旦确定抽样方案，就要求从现场指定批次的钢筋中选取出相应数量的试验试件，将它们送到试验室进行试验。

检测试验一般包括抗拉强度试验、屈服强度试验和疲劳试验。相关的国际标准和法规以及当地的规范会明确指明如何进行这些试验，并对试验结果进行解释说明。钢筋具体需要进行哪些类型的试验取决于项目所处地区及该项目规定应符合的标准和法规。一般情况下，从这些标准和法规中可以获得如下信息：使用的设备类型、对应关键词汇和符号的解释、准备试样的方法、试验的步骤及试验的预测结果。

收到试验报告后，应由经过授权的专业工程人员对其进行解释和说明，并决定下一步的工作内容。若试验结果满足标准和规范的要求，则可以进行后续的施工活动；若试验结果不理想，则需要从同批次的钢筋中再次进行抽样和试验，关于第二次试验所需要抽样的数量，应当参照相关的标准和法规的要求并听取工程师的意见。

进行钢筋试验的试验机构应由业主和咨询工程师共同指定，同时试验报告应提交给他们来进行决策。此外，应存档保存试验报告。

2. Reading：Placement of Rebar（钢筋铺设）

（1）Introduction

There are some marks on the rebar，these marks say where the rebar should be placed and tied, it is necessary to check this information before placing it. Here marks and position of rebars will be discussed.

(2) Text

There are two marking systems in use today: one method is that all bars in one type of member are labeled with the mark of that member; this marking system is for beam rebars, footing rebars, and column rebars. However, the other method is to label the bars with more details. In this case, the marks on rebars inform ironworkers where exactly the bar is to be placed since marks include information on the type of member, floor, size, and the individual number of each particular rebar. With marks on the rebars, ironworkers know where the rebars should be placed while inspectors can check whether or not the rebars fit where they shall. Marks on rebars shall be in their language.

The number of rebars shall be decided according to local standards. Once it is calculated, the size and spacing of rebars must be chosen. Usually, rebars are placed in two layers or four layers, while satisfying the requirement on thickness of cover according to the different laying positions. On construction site, ironworker place the rebars according to the drawings which show the grade of rebars, length of bars, bends, and positions. Rebars shall be placed in the exact right positions as practical as possible and secured well enough. Marks on rebars also help ironworkers to place rebars.

The size and grade of rebars shall be checked before placing them inside the formwork.

The column on the foundation and wall dowels shall go down into the bottom mat of reinforcement where the hooks of dowels can function as support and elevation control. Of course, the dowels must be tied at the bottom and top mats of reinforcement to provide some stability.

It is good practice to place the typical stock length in the middle of the reinforcing rebar run while the short rebars at both ends of the rebar run. It is also recommended to have as few rebar laps as possible.

In short, rebars shall be placed according to drawings at the right locations with sufficient clear distance among parallel rebars in beams, columns, footing, and other elements.

The spacing of rebars in the interior layers and outside layers shall have the same patterns so that the concrete can be placed without any obstruction. It is better to have a smaller size of rebars in the interior layers than in the outside layers. A clear spacing of at least 3 in. between rebars themselves shall be allowed to facilitate the pouring of the concrete.

Sometimes an opening shall be allowed on the top of the reinforcement to allow the ironworkers into the cage, while the rebars of the opening be positioned before concreting.

Additional rebars are needed if there are some changes with substructure support conditions. They shall be placed both in the interior and outside layers. The spacing of rebars shall be multiple or submultiple of the spacing of the main reinforcement.

Once rebars are placed at exact positions with desired spacing between them, they must be secured with tie wires.

(3) Words and Phrases

mark 标识
ironworker 钢筋工，绑扎钢筋工
dowel 插筋，纵筋
mat of reinforcement 钢筋网
sump pit 集水池
interior layer rebar 内层钢筋
outside layer rebar 外层钢筋

(1) 导读

钢筋上通常有一些标识，这些标识会说明钢筋的铺设位置和绑扎方式，因此在铺设钢筋前有必要核对一下这些信息。本篇将着重介绍钢筋的标识和铺设。

(2) 正文

目前施工现场使用的钢筋标识有两种。其中一种标识系统中，同一种结构构件的所有钢筋均用该种构件的标识来进行标注，这种标识系统通常用在梁钢筋、基础钢筋和柱钢筋上面。另一种标识系统在标注时会带有更多的信息，例如用这种系统标识的钢筋能够告诉钢筋工具体某一个钢筋应在何处铺设，因为标识包括了相关信息，如结构构件的类型、楼层、钢筋尺寸及顺序编号。通过查看钢筋上面的标识，钢筋工就可以知道某一个钢筋的铺设位置，同时质检人员也能够检查出该钢筋的位置是否符合图纸的要求。此外，钢筋的标

识应使用钢筋工的母语。

钢筋的数量应按照当地的行业标准来确定。一旦钢筋的数量被确定，它们的尺寸和间距也可被相应地计算出来。通常来讲，钢筋被设置为两层或四层，按照环境类别满足钢筋保护层的厚度要求。在施工现场，钢筋工应该按照施工图纸来铺设钢筋，施工图纸上标注有钢筋的规格、长度、弯曲角度和铺设位置。应最大可能地将钢筋铺设在规定的位置并固定好，钢筋上的标识也会帮助钢筋工铺设钢筋。

在模板系统内铺设钢筋之前，应首先检查一下它们的直径和级别。

基础柱和墙体的纵筋必须一直延伸到底部的钢筋网，这样纵筋的弯钩可以起到支撑的作用并用来调整标高。当然，纵筋在底部和顶部的钢筋网上都需要被绑扎固定好，以便起到一定的稳定作用。

值得推荐的一个做法是将整根钢筋铺设在纵向或横向钢筋总长的中间部分，同时把短钢筋铺设在两端。对于搭接的数量，建议越少越好。

简单来讲，应该按照图纸的要求将钢筋铺设在正确的位置上，并且保证在梁、柱、基础和其他结构构件中平行铺设的钢筋之间保留有足够的间距。

内部钢筋层和外部钢筋层中的钢筋间距应保持一致，这样混凝土浇筑就不会遇到相应的阻碍，此外最好内部钢筋层的钢筋直径比外部的小一些。钢筋和钢筋之间至少应保留 3 英寸的间距，以便顺利浇筑混凝土。

有时，为了方便钢筋工进出钢筋笼，可在顶层钢筋层预留出一个孔洞，在浇筑前对预留孔洞的钢筋进行铺设。

若底层结构构件的支承条件发生了一些变化，则应该增设一些钢筋，这些增设的钢筋应位于内部钢筋层和外部钢筋层，同时铺设钢筋的间距应是主钢筋间距的倍数或约数。

一旦钢筋被铺设在图纸规定的位置上且钢筋之间的间距符合相应要求，就应将它们绑扎固定牢靠。

3. Reading：Rebar Ties and Supports （钢筋支撑绑扎）

（1） Introduction

How to tie and support rebars properly has something to do with the quality of concrete works，here methods of rebars being tied will be discussed.

（2） Text

To prevent the displacement of rebars during concreting，they are secured with tie wires. Those tie wires come in different sizes，typically soft and annealed. If the rebars are epoxy-coated，PVC ties are required to be used.

First, choose the method to tie the rebars, either a bulk roll or a bag tie spinner. Second, select the suitable type of tie which includes a cross tie, a saddle tie (as in Fig. 5-1), and a combination of a cross and a saddle tie. Lastly, tie these ties with pliers.

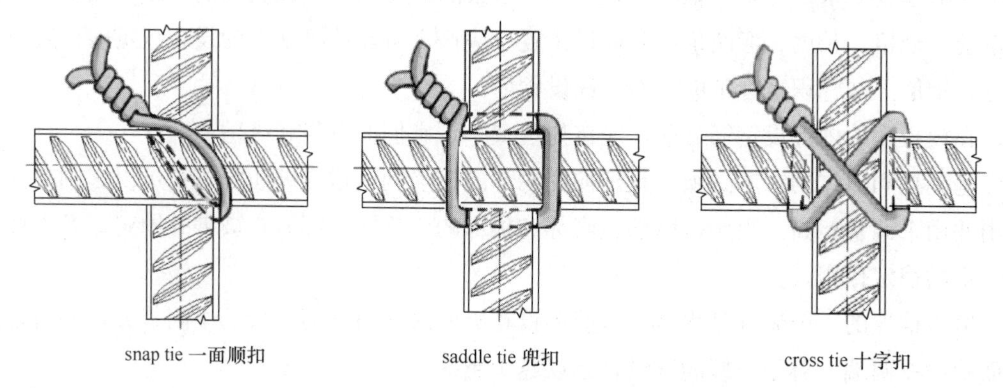

Fig. 5-1 Types of Rebar Ties 钢筋绑扎类型

It is not necessary to tie every intersection of rebars, except those intersections subject to forces from two directions. The ties will not increase the strength of a structure, the reason why tie wires are there is to prevent the displacement of rebars during the concreting. That is all. But be careful that the ends of the tie wires shall be kept away from the surface of concrete elements.

Stirrups shall be tied to the main rebars with snap ties, while it is best to put the stirrups and main rebars together outside the form and then place them inside the form as one unit.

To cast high-quality concrete work, rebars shall be tied sufficiently, apart from being supported with the right kind of supports, like chairs, concrete blocks, and others.

Positioning rebars at the right heights is critical for the structural stability since a little bit higher bottom rebars or lower top rebars would decrease the load-carrying capability. Therefore, it is common practice to place the right height of rebar supports under the rebars according to the specifications and drawings. Normally, the rebar supports include steel wires, precast concrete blocks, plastic chairs, and metal supports. What kind of support is chosen depends on the budget of the project.

The placing of rebar supports is decided by the size of rebars for which they support. Usually, it is shown on the construction drawings.

Only rebars are supported by supports firmly and both of them are tied together, could concrete work have the required covers which would protect the rebars in the concrete.

The cover is one of the most important issues to think of when placing the rebars. Between the rebars and the surface of the corresponding concrete element, adequate cover can allow good protection for steel rebars from corrosion and bond between steel rebars and the concrete itself. Generally, the cover is specified in the related drawings. The correct cover is achieved by installing the right height of metal chairs or other supports under rebars and tying them together. Besides, plastic chairs are also used to support the rebars or wire mesh.

Cover of concrete work should be produced by good workmanship together with inspection on time before concreting.

(3) Words and Phrases

tie wire　扎丝
annealed　退火，热处理
epoxy-coated　环氧树脂涂层的
saddle tie　兜扣
stirrup　箍筋
plastic chair　塑料垫块
concrete block　混凝土砌砖，混凝土垫块
cover　保护层
wire mesh　钢丝网，铁丝网

(1) 导读

钢筋绑扎和支撑的好坏将直接关系到混凝土工程的施工质量，本篇将着重介绍钢筋绑扎的方法。

(2) 正文

为了防止钢筋在混凝土浇筑时发生位移，需要用扎丝将它们扎牢。这些扎丝的直径、尺寸各不相同，但通常是比较软且经过热处理的。若钢筋有环氧树脂涂层，则需要使用PVC扎丝。

首先要选择钢筋绑扎的方式，即选择钢丝圈人工绑扎还是绑扎枪绑扎；然后选用合适的绑扎结，绑扎结有十字扣、兜扣（图 5-1）、十字扣和兜扣的混合结这三种；最后用钳子将这些绑扎结拧紧扎牢。

双向受力网交叉点应进行满扎而不得跳扎，绑扎并不能增加结构构件的强度，它只是为了在浇筑混凝土时防止钢筋发生位移，但是一定要注意，扎丝的绑口应朝向里侧，远离混凝土结构构件的表面。

箍筋和纵筋应通过一面顺扣绑扎在一起，最好是将箍筋和纵筋在模板以外的地方先进行绑扎，待绑扎成一个整体后再将其放进模板里。

为了能浇筑出高质量的混凝土工程，钢筋应该被绑扎牢靠。此外，还需要选用合适的支撑方式来支撑钢筋，可选用的支撑有钢筋支架、混凝土垫块以及其他支撑物。

钢筋铺设在设计要求的标高对于混凝土构件的稳定至关重要，因为无论是下部钢筋的标高稍高一些或者是上部钢筋的标高稍低一些，都会降低混凝土构件的承载力。因此，一般的做法是依据图纸和规范要求在钢筋下面放置合适高度的钢筋支撑。通常来讲，钢筋支撑有钢丝支架、预制混凝土垫块、塑料卡和金属垫块，应根据项目的成本预算来选定需要采用的支撑垫块。

钢筋的直径大小将决定其支撑的安放。通常来讲，支撑的摆放位置在施工图纸上会有显示。

钢筋应被支撑牢靠且与其支撑扎紧在一起，只有在这种情况下，混凝土构件才能具有设计图纸所规定的保护层。保护层可用来保护混凝土构件中的钢筋。

铺设钢筋时，保护层厚度是需要认真考虑的一个问题。在钢筋和相应混凝土构件的外表面之间，适宜的保护层可以防止钢筋受到腐蚀，同时确保钢筋和混凝土之间存在粘结。一般来讲，图纸上有对保护层的相应规定和说明。通过在钢筋的下面摆放适宜高度的金属垫块或其他类型的垫块且将它们绑扎在一起，就可以获得设计所要求的保护层。此外，塑料卡也可以用来支撑钢筋或钢筋网。

只有采用精湛的施工工艺，同时在浇筑混凝土前进行及时的检查，才能浇筑出保护层厚度符合设计和规范要求的混凝土工程。

4. Dialogue：Tying Rebar （钢筋绑扎）

At 10 a. m. on Thursday，on one construction site，ironworkers were tying rebars for one slab floor. Jack and David were on the spot，supervising the job.

David：Some ironworkers are using burlap to remove dust and rust from the rebars. Do they do this stuff every time before tying the rebars?

Jack：Yes，they do. They also spray formwork oil on forms before placing the rebars in it，which is preparation for rebar placement.

David：How do they know which rebar should go where?

Jack：Normally，they would consult the rebar placement plan beforehand.

David：What is next after the placement of rebars?

Jack：They would support the rebars with plastic chairs or concrete blocks and then tie them securely and tightly.

David：Any requirement for rebar supports?

Jack：Yeah，supports will determine the thickness of the cover of concrete work，they are shown and described in detail on the construction plan.

David：Several different types of ties，such as snap ties，figure eight ties，and saddle ties，are used with rebars. Should any thoughts be given when it comes to the selected appropriate type of tie?

Jack：Indeed，yes，there is some consideration，like where the rebars are placed and what forces the rebars are exposed to，but the matter here is these rebars shall be tied efficiently with pliers.

David：Will there be shifting of rebars during concreting?

Jack：Usually they will not. Ironworkers and engineers should be there to observe the rebar configuration while the concrete is placed. If it occurs，they would either support the rebars with a handled tool or alter the direction of flowing concrete.

David：I see. Since ironworkers are likely to be injured when working around rebars，Shall any precautions be taken?

Jack：Actually，there are some precautions，like putting caps on the exposed sharp ends of rebars. Alright，let us get closer and have a look.

周四上午10点，在一个施工现场，钢筋工正在绑扎一个楼板钢筋，杰克和戴维同时也在现场进行监督指导工作。

戴维：这几个工人正在用粗麻布清理打扫钢筋上的灰尘和锈迹，每次钢筋绑扎前他们都这样做吗?

杰克：是的。同时，他们会在钢筋放进模板之前，先给模板喷涂一些隔离剂，这些都是铺设钢筋前的准备工作。

戴维：他们怎么知道某根钢筋应该铺设在什么位置呢?

杰克：在开工之前，他们一般会认真查看施工图纸上的钢筋绑扎说明。

戴维：钢筋铺设之后的工序是什么呢?

杰克：钢筋工们会在钢筋下面放置一些塑料卡或者混凝土垫块来支撑它们，并且把钢筋和垫块一起扎紧。

戴维：对于钢筋支撑有什么具体的要求吗？

杰克：有的，因为钢筋支撑决定着混凝土构件保护层的厚度，所以施工图纸上有对它们的详细说明和解释。

戴维：绑扎钢筋有几种不同类型的绑扎方式，如一面顺扣、八字结和兜扣。在选择绑扎方式时需要考虑什么因素呢？

杰克：有一些因素需要考虑，如钢筋的铺设位置，以及它们将会受到的不同的力，但这里需要注意的是这些钢筋都应该用钳子拧紧。

戴维：在浇筑混凝土时，钢筋会发生位移吗？

杰克：在大多数情况下，钢筋不会发生位移。钢筋工和工程师们都会在现场观察浇筑混凝土的情况，以便发现钢筋有无发生任何的位移或者变形。若钢筋有变形或发生了位移，他们会用一些工具来临时支撑住钢筋，或者改变一下混凝土浇筑的方向。

戴维：明白了。钢筋工在绑扎钢筋时很容易受伤，有没有一些可以采取的预防措施？

杰克：是有一些预防措施的，比如在裸露的钢筋头上放置一些钢筋帽。现在让我们走近一些去看一下。

5. Exercises （练习）

(1) Oil on reinforcing bars should be avoided because it reduces the __1__ between the rebars and the concrete.

(2) Rebars are marked to show where they will __2__. You may work according to either one of the two most-used systems for marking bars.

(3) Tie wire is used to __3__ rebar in place to ensure that when concrete is placed the rebars do not shift out of the position.

(4) The proper coverage of rebars in the concrete is very important to protect the __4__ from fire hazards, the possibility of corrosion, and exposure to weather.

(5) To hold the rebars firmly in position, you shall tie the rebars together at frequent __5__ where they cross with a snap.

(6) A stirrup is tied to the main __6__ steel with a snap tie.

(7) A stirrup is a shaped rebar that holds the lateral reinforcement in a certain __7__, often called a cage.

(8) Check your __8__ to make sure each component of the reinforcement is in place.

(9) Once the mat or cage is assembled, you must hold it in __9__ so the concrete will cover it completely.

Unit 6

Concrete (混凝土)

1. Reading：Concrete Manufacturing（混凝土制备）

(1) Introduction

Concrete manufacturing plays an important role in the construction schedule and quality of concrete works. Knowing batching plant and its equipment is the minimum requirement for civil engineers. Here mixer，weighing，liquid-dispensing equipment，and silo/bin will be described.

(2) Text

Concrete may be produced by factory-mixed or truck-mixed，but the concrete used mostly nowadays is mixed at a central batching plant，which consists of a silo/bin，weighing equipment，liquid-dispensing equipment，and mixer.

For silo/bin，it is used to store all maximum-size aggregates，cementitious materials，and liquid admixtures. To preserve those materials stored in it，it shall be able to protect the stored materials from being contaminated and intermixing with each other. In addition，it can discharge the aggregates freely and smoothly in a controlled way. Furthermore，it shall keep those stored materials in a dry condition. For weighing equipment，it shall be equipped with a visual weight indicating device that readout can be seen clearly by the operator. The same is true with liquid-dispensing equipment. A visual metering device shall be able to help the operator control the dispensing of liquid. Lastly，for mixers，there are several types of mixers available：tilting-drum mixer，horizontal-drum mixer，and split-drum mixer，among which the tilting-drum mixer is very popular on today's construction sites. Basically，a mixer accepts batched materials from a central batching plant and mixes materials by rotating the drum at the required speed in one direction on the way to the site. Once the rotation is done well with mixing，it reverses the rotation and discharges the concrete from the drum. All concrete shall be discharged within 90 minutes once mixing begins.

From a quality point of view，records shall be kept which include plant tests on aggregates and sand，calibration certificates for weighing and test equipment，cement certificates，compressive strength test，and slump test.

(3) Words and Phrases

factory-mixed 混凝土搅拌站搅拌

site-mixed 现场搅拌

truck-mixed　混凝土泵车搅拌

batching plant　混凝土搅拌站

silo　筒仓，贮仓

bin　仓室，储存斗

weighing equipment　称重设备

intermix　混合，搅拌，掺合

split-drum mixer　裂筒式搅拌机

tilting-drum mixer　倾翻式搅拌机

horizontal-drum mixer　卧式搅拌机

rotate　旋转

reverse　反向

compressive strength　抗压强度

译文

(1) 导读

混凝土的制备对混凝土结构的施工进度及质量都十分重要，熟悉混凝土搅拌站及其各种具体设备是土建工程师的基本要求。本篇将着重介绍搅拌机、称量设备、水及外加剂的给料设备、骨料堆场和水泥仓库等。

(2) 正文

混凝土的制备可以在搅拌站或混凝土泵车中进行，其中，搅拌站搅拌混凝土是目前应用最广泛的生产方式，搅拌站包括骨料堆场、水泥仓库、称量设备、外加剂的给料配料设备和搅拌机。

对于骨料堆场或水泥仓库来讲，它的用途是储存所有大尺寸的骨料、水泥材料、外加剂。为了保护储存在它里面的材料，骨料堆场或者水泥仓库需防止材料受到污染，且防止材料之间相互掺合。此外，骨料的给料应在受控制且无障碍的情况下顺畅进行，同时这些骨料在骨料堆场内应保持干燥。对于称量设备，它应安装可视化的称量显示仪表，且仪表的读数对操作人员来说应是清晰可见的，这个要求同样适用于水及外加剂的给料配料设备，其可视化计量设备能有助于操作人员控制、调整水及外加剂的添加。最后，对于搅拌机来讲，现场可能有几种类型的搅拌机：倾翻式搅拌机、卧式搅拌机和裂筒式搅拌机，其中倾翻式搅拌机在当今的施工现场比较常见。通常来讲，搅拌机装载来自混凝土搅拌站已经配料完成的材料，在去施工现场的路上以一个方向在规定的转速下通过旋转滚筒来混合搅拌材料，当搅拌混合均匀后反转滚筒，卸料搅拌好的混凝土。一旦混凝土开始搅拌，就应在 90min 内卸料完毕。

从质量控制的角度来讲，涉及混凝土的相关记录应被保存，如骨料和砂子的工厂试验报告、称量和试验设备的校准合格证书、水泥的合格证、抗压强度试验报告和坍落度试验结果。

2. Reading：Concrete Handling and Placing（混凝土运输与泵送）

(1) Introduction

Delivery, distribution, and placement of concrete have something to do with the quality of concrete work, here accessibility to the site, pumping, and compacting of concrete will be discussed.

(2) Text

The concrete shall be transported to the point of delivery on the construction site and distributed to the final location continuously and quickly, without damaging its workability and causing segregation. Therefore, good planning for access to the site, delivery rate, distribution method, and plant, as well as compacting shall be performed and executed according to the best practices.

When it comes to the workability of concrete, it relies on the nature of concrete elements and the project in which it is to be cast. If any workability is lost, some correction must be done according to the nature of the loss. One more thing is some precautions shall be exercised to keep concrete cool and prevent it from drying out. As to segregation, it is a process during which coarse aggregates separate from the mortar. It will cause poor quality concrete such as non-uniform, weak/porous, or honeycombed patches. The way to avoid segregation is to thoroughly mix the concrete and make it cohesive.

In respect of access to the site by concrete pump truck, it shall be made not to delay or interrupt the distribution and placement of concrete. The things that shall be taken care of include road supporting ability and its headroom as well as places to turn and wait.

Concerning delivery rate, it is decided by the accessibility of the concrete pump truck, that is to say, how fast the truck can come and go. Another thing that needs to be thought of is that the delivery rate shall be consistence with the speed at which the concrete can be placed. In connection with distribution methods for concrete, they include chute, barrow, crane, bucket, pump, and pipelines. The methods selected shall be based on the nature of the project, accessibility of the plant, size of the site, and other factors. However, whatever methods are employed, they shall not delay or interrupt the distribution and placement of concrete. For chutes, they are common on sites, although concrete must not freefall more than 3m. As to barrows, they are used on small sites where a small amount of concrete is needed. In terms of crane and bucket, they are used where sufficient time is allowed for the erection of the crane and it is not easy to pump the concrete. Its placing rate is determined by the capacity of the bucket and the speed of the bucket traveling between the final location and pick-up point. With regard to pumps and pipelines, they are the most commonly used methods for distributing concrete, while pumps can be trailer mounted, truck-mounted, or fixed.

Before pumping, a cement slurry or mortar shall be pumped through the pipelines to coat the inner surface of the pump and pipelines. Pipelines shall be supported and secured firmly and appropriately at the right interval along the length. Once pumping starts, it shall be continuous and smooth without any interruption. One more thing is that joints between each section of the pipelines must be watertight.

Concrete shall be placed vertically and at its final position as near as practical. If it is not easy or possible to pour at its right location, a shovel shall be used to move it to the final position. Furthermore, concrete shall be placed in appropriate layers which would allow the needed compaction to happen. The compaction equipment commonly used is immersion vibrators and vibrating beam screeds.

(3) Words and Phrases

segregation 离析
coarse aggregate 粗集料，粗骨料
porous 多孔的
honeycomb 蜂窝状缺陷
chute 滑槽，斜槽
barrow 手推车，独轮车
crane 起重机，升降架
bucket 铲斗
cement slurry 水泥灌浆料，水泥砂浆
watertight 不透水，防水
immersion vibrator 棒式内部振捣器，振捣棒
vibrating beam screed 梁式振捣找平器

译文

（1）导读

混凝土的运输、泵送和浇筑关系到混凝土结构的工程质量，本篇将着重介绍现场道路、混凝土泵送和振捣夯实问题。

（2）正文

混凝土应被运输到施工现场的输送点，并尽快连续地泵送到浇筑地点，同时不能破坏混凝土的和易性，防止其产生离析现象。因此，应依据同行业最佳的施工工艺和实践经验妥善安排好现场的进出路径、混凝土运输的时间和转运次数、泵送方法、混凝土浇筑后的振捣夯实及与搅拌站之间的协调。

混凝土的和易性取决于其所有组成材料的特性及其所应用的项目的特点。如果混凝土的和易性受到破坏，应根据受破坏的程度采取相应的补救措施。另一件需要注意的事情是，应采取预防措施来保护混凝土中的水分，防止其因受热而蒸发。离析是一个粗骨料和水泥泥浆分离的过程，它会导致混凝土出现浇筑质量的问题，如不均匀、强度低、多孔或蜂窝状缺陷。防止离析的方法是充分搅拌混凝土，增强其黏聚性。

混凝土泵车进出施工现场的道路规划应能够保证其不会延误或影响混凝土的泵送和浇筑。道路的承载能力、净高、转弯及需等待的场所等都是要提前加以考虑和安排的。运输的能力取决于混凝土泵车的进出情况，即泵车能以多快的速度进出现场。另一个需要引起重视的问题是，运输能力应和混凝土浇筑的速度保持一致。运输混凝土的辅助设备有滑槽、手推车、吊车、铲斗、泵和管道，具体选择哪一种设备应根据项目的特点、道路、现场规模的大小及其他因素来决定，然而无论使用什么设备，它们都不应延误或影响混凝土的平仓和捣实。滑槽在施工现场使用很普遍，浇筑时混凝土的自然倾落高度不应超过3m。手推车一般在混凝土用量很少的小型施工现场使用。如果施工进度允许，同时现场又不便于使用混凝土泵，安装吊车和铲斗应是首选方案，混凝土浇筑的速度取决于铲斗的大小及铲斗在平仓地点和混凝土供应点之间的运行速度。混凝土泵和输送管是混凝土施工现场另外两种常用的设备，其中混凝土泵通常被安装在拖拉车上、卡车上或者被固定起来。

在正式泵送混凝土之前，应先在输送管里泵送一些水泥砂浆，以便在泵和输送管的内壁上覆一层砂浆。输送管应根据其总长在适当的间隔处进行支撑并固定牢靠。一旦开始泵送混凝土，就应连续不断地泵送下去。此外，各节管道之间的连接头应是防水的。

混凝土应垂直地进行浇筑并尽可能地接近其浇筑点。如果不容易或者不可能浇筑到位的话，应使用铁铲将混凝土推送到其最终的浇筑点。此外，混凝土应适当分层进行浇筑，以便进行必要的振捣夯实，经常使用的振捣夯实设备有振捣棒和梁式振捣找平器。

3. Reading：Concrete Curing （混凝土养护）

(1) Introduction

Curing is the last step and one of the most critical jobs for concrete works, here several different kinds of curing will be discussed.

(2) Text

Curing is a process during which some measures shall be taken to control the loss of moisture after the concrete has been placed in place. It takes time to let the concrete gain strength and durability. Besides, temperature shall be controlled in a reasonable range due to its impact on strength and durability in the long term. As for how long the curing will last, it depends on many factors, such as the properties of concrete, what it is to be used for, temperature and humidity of the working area. The entire curing period of the concrete takes about a month, during which three timeframes shall be kept in mind:

1) 24 to 48 hours -after the initial set, forms of non-load bearing can be removed and people can walk on the surface.

2) 7 days -after partial curing, traffic from vehicles and equipment is allowed.

3) 28 days -at this point, the concrete should be fully cured.

There are different types of curing. One is to leave the formwork there or cover the concrete with a plastic sheet for some time to prevent moisture from evaporating. Another one is to wet the concrete by ponding or spraying water on the surface. Besides, sometimes, curing compounds are brushed or sprayed on the concrete surface to develop an impermeable coating which can decrease or minimize the loss of moisture. However, this curing compound shall fall into the category of plastic sheeting.

No matter what types of curing are employed, some precautions shall be exercised as follows: when plastic sheeting is used, it shall not be contacted with the surface of fresh concrete directly and shall be secured well not to be blown away by the wind. If water is sprayed on the surface, some measures shall be taken to channel the run-off water out of the surface of the concrete. If ponding is used to cure the surface of concrete, the materials for the dam shall not stain the surface of the concrete.

(3) Words and Phrases

curing 养护

curing compound 养护剂
impermeable 不透水，不渗透
run-off water 积水

(1) 导读

混凝土的养护是混凝土工程的最后一道工序也是一项关键的工作，本篇将着重介绍几种不同的养护方法。

(2) 正文

混凝土的养护是指在其浇筑之后所需采取的一些措施，以便控制其水分的流失。混凝土需要经过一段时间才能达到设计所规定的强度和耐久性。此外，温度应该控制在合理的范围之内，从长远的角度讲，温度对混凝土的强度和耐久性有一定的影响。至于养护到底需要多长的时间，这取决于很多因素，如混凝土的特性、应用场合、工作区域的温度和湿度等。混凝土的整个养护过程大概需要一个月，在这期间有三个时间节点需要牢记：

1) 24～48h：在初始凝固以后，非承重侧模板可以拆除，并且允许人员在上面走动。

2) 7d：混凝土部分硬化之后，允许车辆和设备在上面运行。

3) 28d：此时混凝土的养护全部完成。

混凝土的养护方法有很多种，其中一种是先不拆除模板，或在混凝土的表面铺盖一段时间的塑料薄膜来防止水分的蒸发。另一种是沿着混凝土的周围围上一个土埂后蓄上水，或在混凝土的表面喷上水来保证其所要求的湿度。另外在有些时候，可以在混凝土的表面刷上或喷上一些养护剂来形成一层防水涂层，这层防水涂层可以尽量减少水分的流失，这种养护剂的喷刷应属于塑料薄膜的系列。

无论采取哪种养护方法，都应注意一些细节问题，如在使用塑料薄膜时，薄膜不应与刚刚浇筑完的混凝土的表面有任何的直接接触，并且塑料薄膜应被固定牢靠防止被大风吹走。如要在混凝土的表面洒水，则应采取措施保证其表面没有积水。如要在混凝土的周围围上一个土埂来养护混凝土，那么用来堆积土埂的材料不应污染混凝土的表面。

4. Dialogue: Concrete Slump Test (混凝土坍落度试验)

At 9 a. m. on Monday, on one construction site, Jack and David are present when the subcontractor is doing a concrete slump test.

David：Slump is a simple and low-cost test，carried out from batch to batch to check the workability of freshly made concrete. They will do it in a while.

Jack：Any preparation they need to make?

David：Yes，they will clean all the apparatus with water and then apply the oil on the inner surface of the slump cone to make it friction-free.

Jack：How will the samples be taken?

David：Engineer will decide on the sampling according to the codes and specifications.

Jack：Any particular requirements for a slump cone?

David：Both ends of the cone are open. The diameter of the base end is much bigger than the one of the top end.

Jack：How about the tamping rod?

David：It is steel and a long one with bullet-nosed.

Jack：Any other tools they need for the test?

David：They also need a scoop，base plate，tape measure，and brush.

Jack：Are they starting now?

David：Clearly，they are placing all the apparatus together.

Jack：Can we get closer? I would love to see it.

David：Of course，we can.

Jack：One labor is standing on the footholds of the cone，why is that?

David：He is securing the slump cone to the base plate. You know，the cone must remain still throughout the test.

Jack：They are filling the cone?

David：Yes，they are，first，they scoop some samplings to fill 1/3 of the cone.

Jack：They are tamping now，how long will they do for?

David：Rod 25 times on the first layer，covering the entire cross-section of the sample.

Jack：Refilling the sample again?

David：They fill the cone to 2/3 full for the second layer.

Jack：Rod same times as before?

David：Certainly，yes，but the rod needs to penetrate the first layer by about 1 inch.

Jack：Any pattern on rodding?

David：They would rod circularly，working to the center.

Jack：They are filling for the final layer now?

David：Yeah，they will fill to slightly overflowing，then rod again，also 25 times.

Jack：What is the next procedure?

David：They will strike off the excess concrete from the top of the cone，using the tamping rod.

Jack：Will they be cleaning excess concrete overflow from the base of the cone?

David：Yes，they will. You see，that labor is stepping off the footholds of the cone，holding the come firmly，and lifting the cone slowly and vertically.

Jack：Where are they going to place the cone?

David：They would put the small side of the cone down，next to the concrete sample on the base plate.

Jack：How will they measure the slump?

David：They place the tamping rod over the cone，with one end suspended over the pile of concrete. Take the tape measure and measure the distance from the bottom of the tamping rod to the displaced center of the slumped concrete. The measurement is the slump.

对话

周一上午 9 点，在一个施工现场，杰克和戴维正在参加分包商工程师进行的一个混凝土坍落度试验。

戴维：坍落度试验简单易做，同时成本又低。它通常用来检查刚搅拌好的混凝土的和易性，且每个批次都要进行试验。过一会儿他们就开始做试验了。

杰克：他们需要准备什么吗?

戴维：他们会用水清洗所有需要用到的工具，接着会在坍落度筒的内壁上涂一些油，使其光滑，不会粘上混凝土。

杰克：坍落度如何进行取样?

戴维：咨询公司的工程师会依据标准规范的要求来决定取样方案。

杰克：对坍落度筒有什么特殊的要求吗?

戴维：筒的两端应为开口，且底部的直径要比顶部的大。

杰克：那振捣棒呢?

戴维：它是一根子弹头型的长不锈钢棒。

杰克：试验还需要别的工具吗?

戴维：还需要铲、拌板、卷尺和刷子。

杰克：他们是不是要开始了?

戴维：显然，他们正在把所需要用到的工具都放在一起，做好准备。

杰克：我们走近些好吗?我想看得清楚一些。

戴维：当然可以。

杰克：为什么一个工人把脚踩在坍落度筒的踏板上?

戴维：他这么做是为了把坍落度筒稳定在拌板上，坍落度筒在整个试验过程中是不能有任何移动的。

杰克：他们正在装填混凝土吗?

戴维：嗯，是的。他们首先会装填混凝土，一直装填到坍落度筒的三分之一处。

杰克：他们需要振捣多长时间？

戴维：第一层需要振捣 25 次，混凝土取样的整个截面都需要振捣到位。

杰克：他们又在装填混凝土了？

戴维：是的，装填第二层混凝土要到坍落度筒的三分之二处。

杰克：是和第一层一样也要振捣 25 次吗？

戴维：是的，但是不仅要在本层振捣，振捣棒还需要插入第一层大约 1 英寸的位置进行振捣。

杰克：振捣有什么具体的要求吗？

戴维：振捣要沿着螺旋线由边缘向中心进行。

杰克：现在他们正在装填最后一层的混凝土取样吗？

戴维：是的，这次会多装填一些，使混凝土稍微高出筒顶。与前两次一样，需要再振捣 25 次。

杰克：下一步的工序是什么？

戴维：他们会用振捣棒把多余的混凝土从坍落度筒顶部刮下去。

杰克：现在他们是在清理坍落度筒底部周围多余的混凝土吗？

戴维：是的，现在那位工人不再踩在踏板上了，他稳稳地拿着坍落度筒，慢慢地、垂直地向上把它提起。

杰克：他准备把它放在什么位置呢？

戴维：他会把它放在混凝土取样的旁边，同时让它的底部朝上。

杰克：如何测量坍落度呢？

戴维：他们会把振捣棒放在筒上，棒的一端延伸至混凝土取样，再拿上一个卷尺，测量一下振捣棒底至坍落的混凝土的中心位置的距离，这个测量值就是坍落度了。

5. Exercises（练习）

(1) Never take a __1__ from the first or last section of the pour, it won't be a true representation of the __2__. The concrete is usually sampled after the 1st cube meter of concrete has been poured to ensure a good sample is taken.

(2) Usually, at least 3 cubes are taken from each sample, so make sure you have taken enough from the pour before it finishes. Do check the specification you are working according to, as sometimes the __3__ of cubes you have to take may vary. The __4__ of sampling shall be identified in client specifications or by the engineer. This could be per batch/load or even per volume poured. Check before you start.

(3) Before the concrete is __5__ into the molds, the molds must be lightly __6__ in a mold release agent. This ensures that the concrete does not stick to the molds and makes it easier to remove the cube.

(4) This concrete is poured into the molds and appropriately tempered so as not to have any __7__. After 24 hours, molds are removed, and test specimens are put in water for __8__. The top surface of these specimens shall be made even and smooth. This is done by placing cement paste and spreading smoothly on the whole area of the specimens.

(5) These specimens are tested by __9__ test machine after seven days of curing or 28 days of curing. The load shall be applied gradually at the rate of 140 kg/cm^2 per minute till the __10__ fail. Load at the failure divided by area of specimen gives the compressive strength of concrete.

Unit 7

Mortar（砂浆）

1. Reading：Mortar Ratio（砂浆混合比）

（1）Introduction

It is natural for sand to have some water in it，and water inside it will affect the ratio of mortar. It is vital to have a mortar with a good ratio of water to cement for the sake of physical appearance and workability of mortar. Here mortar ratio will be discussed.

（2）Text

Sand contains moisture，more or less，and this will have some effect on the quality of the bond of mortar. Furthermore，it is better to measure the sand with a shovel based on experience. How much moisture the sand contains will in turn decide the amount of water needed in the mortar mix，having a moisture consistency of sand is very important. As for sand piles，it is a good practice to cover them with plastic film so that the moisture of sand will not be affected by rain，sun，and wind.

In general，the ratio is 3-parts sand to 1-part cement，although it varies according to the nature of the project and the intended use. For instance，the ratio for a mortar used for a load-bearing wall is different from that for a non-load-bearing wall. But the same ratio for one project must be guaranteed from the beginning to the end to ensure the same shade and strength of mortar all the time throughout the project.

Take care of mixing mortar carefully according to the best practices so that consistency both in performance and appearance may be achieved. In fact，mortar proportions，yield，strength，workability，and color can be guaranteed by consistent measurement of ingredients. Generally speaking，volume proportioning is preferred to weight one.

The bond strength of mortar is decided by workability and water content. Although too much or too little water is not good for the physical performance of mortar，there is no slump test or water-cement ratio for mortar. Hence，whether the content of water is adequate will depend on the feeling of the mason's hand. Based on experience，a mortar with a good ratio will not drop or smear but just cling to the vertical surface of a brick. Therefore，it will make it easy to spread and position the brick. Remember to check the plumb and level of bricks at the same time when laying them.

（3）Words and Phrases

bond　粘结，结合

load bearing wall　承重墙

non-load bearing wall 非承重墙
smear 涂抹，油渍，污点
cling 黏着

（1）导读

砂中会含有一些水分，这是一个自然现象，这些水分会影响砂浆的混合比。合理的水灰比对砂浆十分重要，它关系到砂浆的和易性以及砂浆使用后的外观质量。本篇将着重介绍砂浆的混合比。

（2）正文

砂中或多或少都会含有一些水分，这会对砂浆的粘结质量有一定程度的影响。此外，从经验的角度来讲，最好是用铁铲来测量砂的质量。砂里的含水量会直接影响砂浆里需要掺入的水量，保持砂中的水分含量持续一致是至关重要的。对于砂堆来讲，通常用塑料薄膜将它们盖好，以防止砂里的水分受到风雨和太阳的影响。

通常来讲，砂和水泥的混合比为 3∶1。当然，由于项目的特点和实际应用的场合的不同，混合比也会有相应的变化，比如使用在承重墙的砂浆混合比和使用在非承重墙的就不相同。但是，同一个项目所需要使用的砂浆混合比应从项目开始到结束保持一致，以确保在整个施工期间内砂浆的颜色和强度保持一致。

按照最佳的施工工艺认真仔细地混合搅拌砂浆，能够保证其性能和外观不发生变化。事实上，砂浆中各个成分的含量若能经过定期的计量保持不变，就可以保证其比例、收缩性、强度、和易性及颜色的统一。一般来讲，在体积混合比与质量混合比两种方式中，混合砂浆时更倾向于使用体积混合比的方式。

砂浆的粘合强度取决于其和易性和水分的含量。尽管水分过多或过少会对砂浆的物理性能带来不利的影响，但其不需要进行坍落度试验，也不需要指定水灰比。砂浆中水分的含量是否适宜取决于操作工人的经验。从经验上来讲，混合比例好的砂浆不会掉落或溅污地面，会很好地粘在砖的垂直面上，这样就可以轻松地进行砌块和抹砖。需要记住的是，砌块时，要随时检查砖墙的垂直度和水平度是否符合要求。

2. Reading：Methods of Mixing Mortar （砂浆搅拌方式）

（1）Introduction

Mortar is one of the common materials on construction site and good practice of mixing mortar is critical for its workability and physical appearance. Here how to mix mortar by either hand or machine will be discussed.

（2）Text

The size of the project will decide the method of how to mix mortar, either by hand or by machine. If it is a small project, it is possible and reasonable to use hand mixing to produce the mortar, and if not, machine mixing is preferred.

1）Mix Mortar by Hand

① Prepare a platform or container which could be a wheelbarrow or a wooden board.

② Measure all dry materials, such as cement, sand, and chemicals.

③ First, take half the sand, then the cement and lime, and next the remainder of the sand. Lastly, pull and push them forward and backward on the platform or in the container by using a shovel.

④ Sometimes, shovel the mixture from the bottom to the top and mix them thoroughly. Repeat this procedure until the color of the mixture is even.

⑤ Make a hole in the middle of the mixture and add one or two gallons of water into it, after that shovel the mixture back and forth till a smooth consistency of mortar mixture is achieved.

2）Mix Mortar by Machine

① Put three-fourths of the water, half the sand, and all other ingredients.

② Operate the mixer for a little time.

③ Pour the rest of the sand and water into the drum.

④ Run the mixer for a minimum of five minutes.

⑤ Remember to spend adequate time on mixing, too fewer mixing means nonuniformity and poor workability, while too much mixing would cause segregation of materials and trapping of extra air inside.

Once finishing with mixing mortar, just shovel some mortar to see if they slip off the shovel or not.

Please remember to shovel the mortar back and forth a little after a while, which will add some air to the mortar and give them more workability.

（3）Words and Phrases

wheel barrow 手推车

lime 石灰，氧化钙

shovel 铲，锹，单斗挖土机

译文

(1) 导读

砂浆是施工现场一种常用的材料，精湛的砂浆搅拌工艺对其和易性及物理性能至关重要。本篇将着重介绍砂浆的人工和机器搅拌方法。

(2) 正文

施工项目的规模大小决定了砂浆搅拌的方法，搅拌方法包括人工搅拌和机械搅拌。若项目规模较小，通过人工搅拌来生产砂浆是可行且合理的；若项目规模较大，则更倾向于使用机械设备。

1）人工搅拌

① 准备好一个搅拌平台或者容器，可以是手推车或者木板。

② 测量所有干燥的用来混合的材料，如水泥、砂和添加剂。

③ 用铁铲先取出一半的砂、水泥和石灰，再取出另一半的砂。然后，在平台上或容器里用铁铲前后搅拌。

④ 如果有必要的话，用铁铲将拌合物从底部铲到上面，所有拌合物都需要均匀搅拌，上下前后地反复搅拌铲翻，直至砂浆的颜色均匀。

⑤ 在拌合物的中心处用铁铲挖出一个坑，倒上一或二加仑的水，然后用铁铲前后搅拌，直至砂浆拌合物的稠度均匀。

2）机器搅拌

① 加入四分之三的水、一半的砂和其他的搅拌材料。

② 让搅拌机运行一段时间。

③ 将剩余的砂和水也都倒进搅拌筒里。

④ 让搅拌机运行至少 5min。

⑤ 一定要注意的是，机器搅拌的时间要恰当。搅拌的时间如果太短，砂浆不能保持均匀一致且其和易性较差。相反，如果搅拌时间过长，砂浆的各种材料会发生离析，同时砂浆里会含有多余的空气。

砂浆一旦被搅拌好后，要用铁铲铲出一些以观察砂浆是否会从铲上滑落下去。

注意，在搅拌完成后不久，可以用铁铲前后地搅拌，这样可以在砂浆里添加一些空气，增强砂浆的和易性。

3. Reading: Brick Laying (砌块)

(1) Introduction

Brick masonry is one of the common jobs on construction site, it is necessary to learn a good practice to lay brick. Here the procedure of bricklaying will be discussed.

(2) Text

Before laying brick, it would be better to figure out how many bricks are needed and what thickness of joints is required according to the construction plan.

Weather is very important for brick masonry, so be sure to check the weather forecast for those scheduled days of bricklaying. Do not lay bricks on rainy or snowy days, on rainy days water would get into the mortar and ruin the brickwork, while on snowy days, the mortar would set slowly and crack later after frost enters the mortar.

Brick masonry involves laying brick and installing the rebars, flashing, weeps, metal ties, and anchors. Here are steps on how to lay bricks:

1) Set up profiles and check their plumbness, or use two-story poles with a mark on them.

2) Tie the thread with a pin on the profiles or story poles and pull it tightly. Or tie a thread at the top corner of two bricks which are placed on both ends of the first brick course.

3) Place a dry course of stretcher for the first course, making sure that the edge of the top of the stretcher course runs along the mason's line, but does not touch the line.

4) Pick up the bricks one by one and spread a generous layer of mortar along the length of the wall to make a mortar bed, then score the trowel along the center of mortar by angling the brick trowel so that the tip is pointing downwards at a 45-degree angle.

5) Place first stretcher brick on the mortar bed, and then tap it down gently on its edges and corners two or more times to ensure the level of brick. Pop the level on the top of the brick to see if the bubble of level is in the center or not.

6) Pick another brick and add a little mortar on its one end, then tap it down into the place next to the first one, ensuring that they are aligned along the mason's line. Put the mortar out to the front of the brick by using the tip of a trowel, ensuring the mortar fills all the voids, then scrape off all the excess mortar and collect them with a trowel, then butter the next brick.

7) There shall be a little gap between the top of the brick and the line.

8) Repeat step 6 till the end of the course.

9) Raise the line to the next course, and then check its elevation with the mark on the profiles or story poles.

10) Go on to the next course by putting some mortar on top of the existing course.

11) Place the header of brick in the front, or cut the brick into half and put a half next to the profiles.

12) Pick one brick and place some mortar on its one end, then put it down, tapping gently one or two times on its edges.

13) Repeat step 12 till the end of the second stretcher course.

14) Raise again the line to the next course, and repeat step 12.

15) Lay one course after another till the whole wall is laid up.

It is a good practice to check both vertical and horizontal alignment continuously when laying every course of bricks.

(3) Words and Phrases

bricklaying 砌砖

flashing 挡水，防雨

weep 漏水，渗漏

metal tie 系墙铁

anchor 铰链，固定器

profile 皮数杆，标杆

plumbness 垂直

story pole 皮数杆，楼层标尺

dry course 干底层，干砌底层

stretcher 顺砖

trowel 泥刀，瓦刀

mortar bed 砂浆层

scrape off 刮去，擦掉

butter 沿边将砖铺平在灰浆上

header of brick 露头砖，丁砖

译文

(1) 导读

砌砖是施工现场常见的一项施工活动，有必要学习和了解一些砌砖施工工艺。本篇将

着重介绍砌砖的流程步骤。

（2）正文

在砌砖之前，最好能够依据施工图纸来估算需要使用的砖的数量和灰缝的厚度。

天气对砌砖施工影响较大，一定要提前核对项目计划中需砌砖施工的那些日期的天气情况。避免在雨雪天进行砌砖，因为在雨天砌砖的话，水会浸入砂浆中，进而破坏和溅脏砌砖体；在下雪天砌砖的话，砂浆的凝固会变慢，而且霜渗入砂浆后，不久砂浆会开始崩裂。

砌砖施工项目涉及砌砖和安装拉结钢筋、挡水、漏水板、系墙铁及铰链。下面是砌砖的具体步骤：

1）竖立皮数杆并检查其垂直度，或者使用两个带有刻度的木桩。

2）一种方法是用钉将绳子钉在皮数杆或木桩上，并且将绳子拉紧。另一种方法是在第一皮砖的两端各放两层砖，在两个上层砖的砖角之间系上一根绳子。

3）进行第一皮砖的撂底摆砖，确保顺砖皮的上部边缘沿着准线，但不要接触。

4）沿着墙的方向铺上一层厚厚的砂浆来形成一个砂浆垫层，用泥刀尖向下呈45°沿着砂浆垫层的中心线划出一个沟，然后将砖一块接一块地拿上来。

5）将第一块顺砖摆放在砂浆垫层上，然后在砖的边沿和四角轻轻地向下敲击两次或更多次，以保证砖的平整。将水平仪放在砖的上面，检查水平仪的气泡是否位于中间位置。

6）拿起另一块砖，在它的边上挂上一些砂浆，然后将其紧紧靠着第一块砖轻轻放下，确保它们和准线保持平行。用泥刀尖将砂浆推挤至砖的前面，保证所有的勾缝都用砂浆填满，然后将所有多余的砂浆刮掉，用泥刀收集起来，然后沿着边将下一块砖铺放在砂浆上。

7）在砖的顶面和准线之间应有一些空隙。

8）重复步骤6直至这一皮的最后一块砖完成。

9）将准线提升到下一皮，然后检查其标高是否和皮数杆或木桩上的刻度标识一致。

10）在已经完工的第一皮砖的上面铺填一些砂浆，进行第二皮砖的施工。

11）将丁砖放在前面，或者将一块砖一砍为二，将其中的一半紧紧靠着皮数杆放下。

12）拿起一块砖并在其边上挂一些砂浆，然后放下，沿着其长边轻轻敲击一两次。

13）重复步骤12直至第二皮的最后一块砖完成。

14）将准线提升到第三皮，重复步骤12。

15）像这样一皮砖一皮砖地砌筑下去，直至砌完一堵墙。

在砌每一皮砖时，通常要不断地检查各皮砖的垂直度和水平度。

4. Dialogue：Procedures of Mixing Mortar （砂浆搅拌工序）

At 9 a. m. on Monday, on one apartment building construction site, Jack is learning how to mix mortar from foreman John.

John: We are going to make our mortar from scratch.

Jack: What do we need first?

John: We need Portland cement, fine masonry sand, and clean tap water.

Jack: How shall we do it?

John: We begin by dry mixing the cement and sand, using a separate mixing container for the dry components.

Jack: What is the mixing ratio?

John: The standard mortar mixing ratio is one part Portland cement to three parts sand.

Jack: How to mix them?

John: Add the sand to the concrete and use a mortar trowel to fold them together, till the concrete and sand are completely mixed.

Jack: I find there are some stones in the sand.

John: We will sift the sand through a 1/4-inch wire screen to filter out any bits of stone before adding it.

Jack: How much water do we need?

John: We will consider one part water to three or four parts mixtures as a starting point, adding additional water as needed if the mixture is too dry.

Jack: Do we mix the mortar by hand?

John: No, we use a power drill with a paddle attachment, it will mix the mortar more quickly and easily.

Jack: How to operate the drill?

John: It is easy, run the drill at a low speed until the mix is thoroughly blended.

Jack: Any way to know it is thoroughly blended?

John: No visible water puddles or dry sections remain.

Jack: Anything else to care for?

John: Yes, remember, even if the mortar seems properly blended, we also should keep stirring for at least two additional minutes to ensure a smooth blend.

Jack: What if the mortar seems too dry?

John: Add water as needed in small amounts. It's better to add too little water at a time than too much.

Jack: Is the mortar now ready for use?

John: Not yet, allow the mortar to sit and rest, or "slake", for 10 minutes, immediately after mixing it. After slaking, stir the mortar for about 5 additional minutes.

Jack: How can we know the mortar is mixed properly and ready for use?

John: Yeah, it should have a consistency comparable to thick peanut butter.

Jack: Any chance to know it by experience?

John: Just scoop some of the mortar onto the trowel and tip the trowel 90 degrees. If the mix falls off immediately, it is too thin. If it clings to the trowel, it should be ready.

Jack: How long is mortar workable?

John: Once mixed, the mortar in the bucket should be workable for approximately 90-120 minutes. Avoid mixing more mortar than you can apply in two hours as the remainder will become too dry to use.

周一上午9点，在一个公寓楼的施工现场，杰克正在向施工队队长约翰学习搅拌砂浆的方法。

约翰：我们准备自己来搅拌砂浆。

杰克：需要什么材料呢？

约翰：主要是水泥、细砂和干净的自来水。

杰克：应该如何做呢？

约翰：先把水泥和砂在不加水的情况下搅拌，需要使用一个专门的干燥容器。

杰克：它的混合比是多少呢？

约翰：砂浆的常规混合比是一份水泥配三份砂。

杰克：那如何搅拌呢？

约翰：首先把砂混入水泥中，然后用一个抹刀反复铲翻，直至水泥和砂彻底混合在一起。

杰克：我看见砂中还有一些石子。

约翰：在混入砂之前，我们会使用四分之一英寸孔径的铁筛子来筛一下砂，把其中的石子筛掉。

杰克：需要加入多少水呢？

约翰：开始时，一份水配上三或四份水泥、砂的混合物。若加入水之后的混合物仍较干，可以根据需要适量地再加入一些水。

杰克：我们是人工搅拌砂浆混合物吗？

约翰：不是的，我们会使用一个带叶片的电动搅拌器，它搅拌起来会更快、更容易。

杰克：如何操作电动搅拌器呢？

约翰：这是很容易的，把它设置为低速挡位运行即可，开机搅拌直至完全搅拌均匀。

杰克：我们如何知道它完全搅拌均匀了呢？

约翰：没有出现水泡和干的砂浆块就可以了。

杰克：还有别的事情需要特别注意的吗？

约翰：有的，要记住，即使砂浆表面看似已经搅拌好了，我们还应该再继续搅拌至少2min，以确保搅拌彻底。

杰克：如果砂浆看着太干怎么办？

约翰：可以根据需要适当加上少量的水，每次加水少一些肯定比加多了要好。

杰克：砂浆现在可以使用了吧？

约翰：还不行。在搅拌好之后，还需要让砂浆熟化 10min，放一放。在熟化之后，再把砂浆搅拌 5min。

杰克：那我们如何知道砂浆已经搅拌完成可以使用了呢？

约翰：可以看一下它的稠度，如果稠度看起来类似很浓的花生酱，砂浆搅拌就算可以了。

杰克：这里有什么经验或者诀窍吗？

约翰：只需要铲出一些砂浆到抹刀上，把抹刀立起来，若砂浆立刻滑落掉地，就表明砂浆太稀；若砂浆还能继续粘在抹刀上，那就表示它没有什么问题了。

杰克：搅拌好的砂浆的可用寿命是多长时间？

约翰：砂浆在搅拌好之后 1.5～2h 是可以正常使用的。尽量要避免搅拌太多的砂浆，如果 2h 之内使用不完的话，剩余的砂浆就会变干、变硬，就无法再使用了。

5. Exercises（练习）

(1) Mortar is one kind of building material composed of __1__, which in this case is mixed with fine __2__ and water, with lime added to improve the durability of the product. Adding water to this mix activates the cement so that it __3__, or cures, just as with concrete.

(2) Mortar is commonly sold in bags, in a __4__ pre-mixed form that you combine with water. It can also be mixed on-site, using a cement __5__ or simply mixing with a shovel or hoe in a wheelbarrow or mixing tub.

(3) Grout is a similar product that can be seen as a form of mortar but formulated without the lime __6__. Mortar has a higher water content to allow it to flow and fill __7__ between ceramic and stone tiles. Because of its higher water content, grout is not a __8__ material but serves merely to fill gaps.

(4) Thin-set is a related product made of cement and very fine sands, along with a water-__9__ agent such as an alkyl derivative of cellulose. It is used to __10__ ceramic and stone tile to a substrate, such as cement board.

Unit 8

Dampness and Its Prevention

（潮湿及其防护）

1. Reading：Causes of Dampness（潮湿的原因）

（1）Introduction

Dampness can be caused by various factors，such as moisture from the ground，rain，and condensation of air in the buildings. Unfortunately，it harms people and things in the buildings. Here the causes of dampness will be discussed.

（2）Text

Generally speaking，dampness is caused by moisture rising from the ground，rain splashing，exposed top of the parapet wall，leakage from pipes，condensation of air，and defects in the construction and designing of buildings.

For moisture from the ground，since the foundation of all buildings rests on the soil which contains moisture of water or is near groundwater，while groundwater table can rise sometimes during raining season. When moisture rises up by capillary action to the foundation，it would be absorbed by flooring or wall materials.

With respect to rainwater，it can penetrate the external walls by rain splashing if walls are not plastered properly，go down into the wall and roof if downtake pipes are not installed according to the requirement of codes，enter into the slab resulting from water ponding on the flat roof with poor sloping.

On the subject of exposed top of the parapet wall，water can enter the parapet wall if the top is not treated well with dampness proofing course.

Concerning leakage from pipes，it is another potential source of dampness，as a lot of pipes run within buildings，connecting with water tanks，taps，sinks，toilets，tubs，and showers in the kitchen and bathroom. All these pieces of pipes must be connected and joined with each other through different fittings to extend or make a turn. It is those joints where leakage occurs that causes dampness in the buildings.

With regard to condensation of air，due to the improper orientation of a building，there is not enough sunshine in the buildings where humid air exists，so humid air can not be ventilated very well.

In the matter of defects in the construction and designing of buildings，the design of flat roof does not meet the requirement of proper sloping，and poor workmanship of draining system in the buildings has been done.

In brief, one cause or another or a combination of them can result in dampness in the buildings which brings some bad effects on the people and objects in the buildings.

(3) Words and Phrases

dampness　潮湿，含水率
splash　溅湿，溅污
parapet wall　矮护墙，女儿墙
capillary　毛细管作用
penetrate　渗透
plaster　灰浆，抹灰，涂灰泥
dampness proofing course　防湿层
fitting　接头，配件，零件
ventilate　通风

(1) 导读

造成建筑物潮湿的原因可能有很多种，比如土壤中的水分、雨水和建筑物内空气水分的凝聚等。然而，潮湿会对在建筑物中居住的人和存放的物品有不良的影响。本篇将着重介绍引起建筑物潮湿的各种原因。

(2) 正文

通常来讲，造成潮湿的原因有：地下土壤中所含水分的蒸发、雨水的溅淋、女儿墙外露的顶部及管道的泄漏、空气水分遇冷后的凝聚以及楼房设计建造过程中存在的缺陷。

关于地下土壤中的水分蒸发这一原因，所有建筑物的基础都建造在土壤之上，而土壤中又含有水分，或者接近于地下水，在雨季时地下水位有时会上升，当水分由于毛细管作用上升到基础地面时，它就会被地板或墙面的材料吸收。

至于雨水，若外墙涂抹泥灰的工艺不符合规范要求，下雨时雨水则会溅到外墙上并随后渗透进外墙里；若雨水管没有按照规范要求进行施工安装，雨水就会流到墙和屋顶上；若平顶屋顶的坡度不符合设计要求，则屋顶上就会形成积水，进而渗透到楼板里。

对于女儿墙外露的顶部来说，若没有按照设计要求对其进行适当的防潮处理，雨水就会进入女儿墙里。

管道泄漏是另外一个潜在的潮湿原因，因为房屋里面都安装有大量的管道，这些管道连接着厨房以及卫生间里的水箱、水龙头、地漏、厕所、浴缸和淋浴头。所有这些管道必须通过不同的管件配件相互连接在一起，或是延伸管道，或是拐弯一个角度，在这些管件连接的地方容易发生泄漏，从而导致房屋出现潮湿的情况。

关于空气水分的凝聚，由于房屋的朝向不好、室内潮湿或者屋里的光照不充足，潮湿空气无法很好地通过通风排出去。

最后，楼房设计建造过程中存在的缺陷包括平顶屋顶的设计不满足适度坡度的要求，或者房屋雨排水系统的施工方法不符合规范要求。

总之，各种原因以及它们的组合导致了房屋出现潮湿的问题，而潮湿问题又进一步给房屋里的居民和物品带来一些不良的影响。

2. Reading：Dampness Damage and Its Prevention （潮湿的危害和预防）

(1) Introduction

Dampness can harm elements of structure, things, and people in the buildings, and it is necessary to apply dampness courses in and on the buildings. Here the prevention of dampness will be discussed.

(2) Text

Dampness has a bad effect on elements of structures, objects in the buildings, and people working and living in buildings. Those effects follow into several categories:

1) Patches emerge here and there on the walls, or plaster on the wall becomes soft and crumbles, or the color of wallpaper changes, or a wallpaper breaks apart.

2) Patches appear on ceilings, or some parts of a ceiling crumble and fall during raining season.

3) Wooden floorings warp and carpets decay.

4) Wooden furniture deforms and could not function well.

5) Metal furniture or objects buried in the buildings might be corroded.

6) Electrical appliances can be short-circuited due to the dampness in some worst cases.

7) Dampness provides a suitable environment for breeding and growth of mosquitoes and termites which can cause diseases to people in the buildings.

Therefore, it is necessary to choose materials that meet the requirement of damp proofing and apply them properly to prevent dampness in the buildings.

There are many kinds of materials that can be used for preventing dampness in buildings. They include hot bitumen, mastic asphalt, bituminous felt, plastic sheet, metal sheet, a combination of lead sheet and bituminous felt, engineering brick, mortar with chemical additions, and concrete with dampness prevention chemicals.

First, bitumen is in a state of fluid when heated and can be applied on concrete bedding in a pass of 3mm thick. Second, mastic asphalt is a combination of heated asphalt, sand, and minerals which shall be applied according to specification from the manufacturer. Third, bituminous felt comes in a roll and can be laid on roof slabs with the help of torches. The three materials mentioned above are the most common materials used nowadays, which are applied on elements of buildings in different ways based on the nature of the elements.

(3) Words and Phrases

patch 碎片，斑片
crumble 破裂，破碎
warp 扭曲，弯曲
decay 腐烂，腐坏
corrode 腐蚀，侵蚀
mosquito 蚊子
termite 白蚁
bituminous felt 沥青油毡

(1) 导读

潮湿会对各种建筑结构和建筑物中的居民及物品产生不良的影响，因此十分有必要对建筑物进行相应的防潮处理，本篇将着重介绍潮湿的预防。

(2) 正文

潮湿会对建筑物的结构、建筑物里的物品和在其中工作生活的人们产生不良影响。这些不良影响包括：

1) 墙体上的不同部位会出现斑点，墙上抹的灰泥会变软破裂，墙纸的颜色可能会改变，还有些墙纸会破裂。

2) 屋顶上会出现斑点，在雨季，有些屋顶会部分破裂并掉落。

3）木地板会弯曲变形，地毯会腐烂。

4）木制家具会变形，以至于无法正常使用。

5）金属制造的家具或者预埋在房屋里的金属构件会受到腐蚀。

6）在一些极端恶劣的情况下，家用电器会因为潮湿而发生短路。

7）潮湿为蚊子和白蚁的生长提供了一个适宜的环境，而这些蚊子和白蚁会给在房屋里居住和生活的人们带来疾病。

因此，必须选取满足防潮要求的卷材和涂料并正确喷涂，以预防房屋的潮湿问题。

市场上有很多种材料可用于房屋的防潮，如热沥青、沥青砂浆、沥青油毡、塑料膜、金属板材、铅板和沥青油毡的合成板材、工程抗蚀砖、含化学添加剂的砂浆和防水混凝土。

沥青加热后是一种流体状物质，可以涂刷在混凝土的基层上，每层的厚度为 3mm；沥青砂浆是一种混有加热的沥青、砂和矿物质的合成材料，其应按照厂家的使用说明进行涂刷；沥青油毡成卷出厂，可以借助汽油喷灯将其铺设在屋顶的楼板上。上面所说的三种材料是当今使用最为广泛的防潮材料，可以根据各建筑构件的特征，使用不同的方法将它们铺抹喷涂在建筑构件的表面。

3. Reading：Dampness Proofing （防潮）

(1) Introduction

Dampness Proof Course (DPC) is one of the common methods to prevent dampness in the building and DPC should be applied continuously. Here DPC and cavity wall will be discussed.

(2) Text

Many methods can be taken to prevent dampness in the structures, among which applying DPC (as in Fig. 8-1) and construction of cavity walls are very popular. Appling the DPC course means that DPC materials are brushed or laid on the leveled bedding surface or surface of the wall between the source of dampness and building elements, which shall cover the whole width of the wall. The construction of cavity walls can prevent rainwater from penetrating the inner wall of buildings.

Furthermore, the DPC course shall be continuous anywhere, including joints and corners. If felts are applied, overlay between two neighboring felts shall meet the requirement of codes and be sealed according to the instruction from the manufacturer.

Dampness is a big problem that needs to be taken care of seriously to protect the ele-

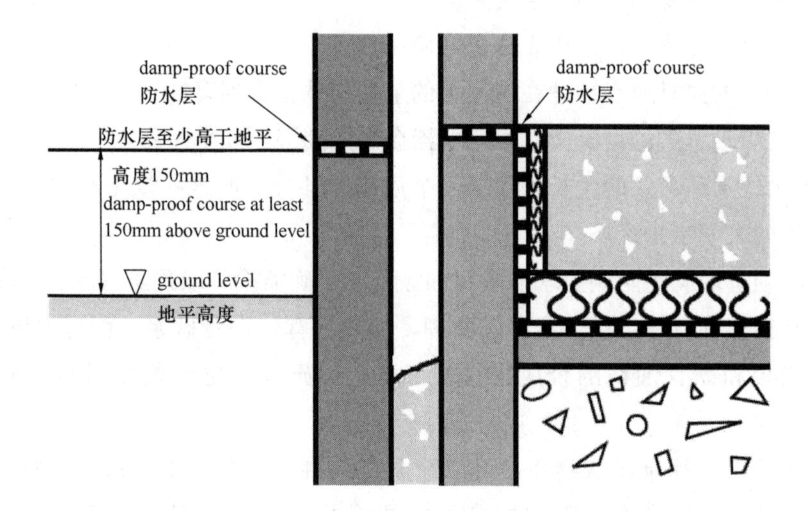

Fig. 8-1 Damp-proof Course 防水层

ments of buildings and the people in the buildings. There are so many materials that can be applied in different ways to prevent dampness in the structures.

(3) Words and Phrases

cavity wall 空心墙，双层壁

overlay 在……上铺涂，重叠，涂层

(1) 导读

安装防水层是预防房屋潮湿的一种常见方法，防水层应该连续铺抹。本篇将着重介绍防水层和空心墙。

(2) 正文

预防建筑物潮湿的方法有多种，其中铺抹防水层（图 8-1）和建造空心墙这两种方法比较常见。铺抹防水层是指将防水层材料涂刷在平整后的基底表面或潮湿源和建筑结构件之间的墙的表面，并且整个墙面都要涂刷。建造空心墙可以防止雨水穿透进入建筑物的内墙。

此外，必须连续地涂刷防水层，并且要涂刷接缝和阴阳角。若使用油毡，两块相邻的油毡之间的搭接缝应符合规范要求，并且应根据厂家的说明对搭接缝进行密封处理。

潮湿问题是房屋建筑中一个非常严重的问题，需要认真对待和处理，这样才能进一步保护房屋里的人员和物品。市场上有多种材料，可以通过不同的涂刷方法来防止建筑构件

出现潮湿。

4. Dialogue：Waterproofing in Toilet （卫生间防潮处理）

At 9 a. m. on Tuesday，on one apartment building construction site，Jack is learning the know-how of waterproofing for the toilet from foreman John.

John：First，let us prepare the floor surface of the toilet for waterproof built-up.

Jack：As always，we will clear dirt，dust，soft mortar，and all loosely adhered particles，won't we?

John：Quite right. Then we are going to do a water ponding test.

Jack：How shall we do it?

John：We are going to have a water pond，let water stand on the floor of the toilet for 48 hours.

Jack：What shall we look into?

John：If water is running short，it tells us that there are some leakages.

Jack：Leakages are caused by some cracks. If this is the case，are we going to repair them?

John：Of course，we will create some V grooves，and fill them with grouting.

Jack：How about those pipe inserts?

John：We will apply some mix of cement and sand on the openings on the walls around the pipes.

Jack：Are we ready to apply the base coat now?

John：Not yet，the slab must be pre-wetted，no running water on the slab.

Jack：That is what we bring a dry cloth for?

John：Yeah，then we will apply the first coat of cement slurry with a waterproofing compound.

Jack：Will it cover the whole floor?

John：Yes and no，the coat also has to be extended on the masonry wall at least 2 inches above the sunk slab.

Jack：Shall we leave it to dry for a while?

John：Definitely，at least 6 hours before applying the second coat.

周二上午9点，在一个公寓楼的施工现场中，杰克正在向施工队队长约翰学习如何做卫生间地面的防水工作。

约翰：首先，我们来做一下准备工作，也就是卫生间地面防水施工前的准备。

杰克：我们和往常一样先清理地面，清除尘土、砂浆软块和其他粘在地面上的东西，对吗？

约翰：对的，接着我们做蓄水试验。

杰克：蓄水试验如何做呢？

约翰：我们会在卫生间的地面上蓄水，保持48h。

杰克：我们要观察什么指标呢？

约翰：如果蓄水变少了，就表明卫生间的地面存在泄漏点。

杰克：泄漏是由裂缝引起的，如果真有裂缝的话，我们是不是需要对它们进行修补？

约翰：当然，我们会在地面上凿出一些V形槽，然后用砂浆灰泥把它们填充饱满。

杰克：对那些管根和孔洞如何处理呢？

约翰：我们会提前搅拌出一些水泥砂浆，把它们抹在墙上所有的管根和孔洞的四周，并且抹实、抹满。

杰克：现在可以做第一层的防水了吧？

约翰：还不行。地面上需要提前洒一些水来让它湿润，而且地面上不能有任何多余的明水。

杰克：我们带来干抹布就是用来擦干地面上的水吧？

约翰：是的。下面我们做第一层的防水，涂刷由水泥和防水添加剂混合而成的防水层涂料。

杰克：是整个地面都要做吗？

约翰：不仅仅是地面，防水工作同时还需要在墙上至少做2英寸高。

杰克：做完防水的地面是否需要干燥一段时间呢？

约翰：当然。在做第二层的防水之前，第一层的防水层至少要干燥6h。

5. Exercises（练习）

(1) There are ___1___ or membranes of water ___2___ materials such as bituminous felts, mastic asphalt, plastic ___3___, cement concrete, mortar, metal sheets, stones, etc. which are interposed in the building structure at all locations wherever water ___4___ is anticipated or suspected.

(2) The surface treatment consists in filing up the pores of the material exposed to ___5___ by providing a thin ___6___ of water repellent material over the surface (internal/external). External treatment is effective in preventing dampness.

(3) The integral treatment consists of adding certain ___7___ to the concrete or mortar during the process of ___8___, which, when used in construction, acts as barriers to moisture penetration under different principles.

（4）A cavity wall consists of two parallel ___9___ or leaves or skins of masonry separated by a continuous air space or ___10___. The provision of a continuous cavity in the wall effectively prevents the transmission or percolation of dampness from outer walls or leaf to inner wall or leaf.

Unit 9

Plastering（抹灰）

1. Reading：Wall Plastering Preparation（抹灰准备）

（1）Introduction

Plastering is vital for the physical appearance of a building，so it is important to prepare for it，such as clean tools，freshly mixed plaster，and drop cloths. Here preparation before formal plastering is discussed.

（2）Text

Plastering on both interior and exterior wall is one of the final steps of a construction building，also one of the most important jobs due to its effect on the appearance of a building. Here we are going to cover the materials and tools used to do plastering and how to prepare for plastering the walls.

Freshly mixed plaster and PVA glue are needed in working places，along with tools，such as a trowel，hand board，bucket，mixer，and a water brush. In addition to those above，some pieces of plaster sheet and drop cloth should be available. In a word，clean tools and freshly mixed plaster must be handy before preparing the wall.

Preparing the wall poorly will affect the connection between plaster and wall，in the end，the quality of the plastering job. Preparation of wall usually consists of covering the floor，electrical sockets，and fittings with drop cloths；clearing any grime，oil tar，and debris on the wall；washing the wall with a sufficient amount of water，applying scrim tape for gaps between plasterboards；rolling PVA glue on the wall. If one of those tasks is not performed well，plaster cannot establish a good bonding with the wall or make a mess on other works. For example，without properly applying PVA on the wall，it is more likely for the plaster to form budge，or fall completely.

Only when working surface is prepared thoroughly and well can plastering begin.

（3）Words and Phrases

plastering　抹灰
trowel　泥刀，抹子
hand board　泥浆托板
bucket　水桶，桶状物，铲斗
drop cloth　罩布
socket　插座，套节

grime 尘埃，积垢
oil tar 焦油
budge 移动，变化

(1) 导读

室内和室外的抹灰对房屋的外观质量十分重要。因此，抹灰之前的准备工作，例如准备清洁的工具、刚搅拌好的泥浆和罩布，就显得非常有必要。本篇将着重介绍正式抹灰前的准备工作。

(2) 正文

室内室外的抹灰是房屋建筑施工最后工序中的其中一个，也是其中一项重要的施工活动，它涉及房屋的外观质量。本篇将着重介绍一下抹灰所需要的材料、工具以及抹灰前的准备。

工作现场除了必需的工具，比如泥刀、泥浆托板、水桶、搅拌器和刷子之外，还需要准备刚刚搅拌好的泥浆和 PVA 胶，此外还需要准备一些塑料单和罩布。总之，在墙体抹灰之前，清洁的工具和刚搅拌好的泥浆应保证随时能够使用。

墙体的准备工作如不符合规范要求，将会影响泥浆和墙体基层之间的粘结强度，最终会影响整个抹灰工序的施工质量。墙体的准备工作通常包括：用罩布遮盖住地板、电源插座和配件等；清除墙体上留下的灰土、油渍和碎屑；用足够量的清水清洗墙体；在石膏板之间的缝隙粘上自粘网络胶带；在墙体上滚涂 PVA 胶。如果上述这些施工工序中的任何一项没有做到符合规范要求，泥浆就不会和墙体形成良好的粘结，或会给其他工作带来不利的影响。例如，若墙体上的 PVA 胶的滚涂方法不得当，则泥浆极有可能形成空鼓，或者整个墙体的泥浆可能脱落到地面上。

只有在整个墙面都准备完成之后，才能开始抹灰。

2. Reading：Procedures of Mixing Plaster （泥浆搅拌）

(1) Introduction

Clean tools are essential for plaster mixing, which should be done properly according to the best practice and codes. Here tools needed for mixing plaster and mixing procedures will be discussed.

(2) Text

Firstly, prepare tools before mixing plaster. Tools include a bucket, trowel, water brush, and mixer with a paddle, among which trowel is used to scrape any lumps or dry plaster off the inside surface of the bucket. As for the water brush, it is used to clean the paddle of the mixer and trowel once mixing is done.

Secondly, get water and plaster ready on site. Water must be fresh and clean, while plaster shall not be expired according to the label on its bag.

Thirdly, it comes to the procedures of mixing plaster :

1) Pour half a bucket of water into a thoroughly clean bucket.

2) Add approximately half of the total amount of plaster into the water for the first time.

3) Mix the plaster immediately by swirling it in two directions, clockwise and anti-clockwise separately.

4) Add approximately 50% of the remaining amount of plaster into the water and mix them.

5) Scrape round the bucket and mix any lumps again.

6) Scoop some plaster on the trowel to see if any plaster is sliding off the flat trowel. If yes, add another small amount of plaster and mix them.

7) Check whether the plaster is free of lumps, smooth, and creamy or not. Or put a stick in the mix to see if it can stand upright on its own, if yes, plaster is mixed well enough.

Lastly, wash the bucket, trowel, and paddle as soon as mixing is done, otherwise, plaster left on the tools will dry on them, causing further problems with future plastering.

(3) Words and Phrases

paddle 桨，叶轮
lump 团块，隆起
swirl 旋转，旋动，弯曲
scoop 用铲取出，用勺取出

(1) 导读

清洁的工具对泥浆的搅拌效果有很大影响，且泥浆的搅拌须按照规范的要求和同行业的最佳工艺来操作。本篇将着重介绍泥浆搅拌所需要的工具和其具体的流程。

（2）正文

第一，在搅拌泥浆之前需要准备相应的工具。这些工具包括水桶、泥刀、刷子和带叶轮的搅拌器，其中，泥刀用来刮掉附着在水桶内壁的任何泥浆块，刷子则主要用来在泥浆搅拌完成之后清洁搅拌器的叶轮和泥刀。

第二，水和泥浆在施工现场应提前准备好。水应是干净的流动水，泥浆应是没有过期的。

第三，泥浆搅拌的步骤如下：

1）在完全清洁干净的水桶里倒入半桶的水。

2）第一次先加入大约一半量的干泥浆。

3）立即分别按顺时针、逆时针两个方向来搅拌混合泥浆。

4）再次在水中加入剩下的大约一半量的干泥浆并继续进行搅拌。

5）刮一下水桶的内壁，再将出现的团块混合搅拌。

6）铲出一些泥浆到泥刀上，观察是否有泥浆从平放的泥刀上滑落。若有泥浆滑落，则在泥浆中再加入一些干泥浆并继续搅拌。

7）检查泥浆里是否还有团块存在、泥浆是否平滑黏稠，或者也可以在泥浆中放一根棍子观察其是否能够直立，若其能直立，则表明泥浆已经搅拌混合得充分且彻底。

最后需要注意的是，混合搅拌一旦完成，需要立即清洗水桶、泥刀和叶轮，否则残留在工具上的泥浆会干燥成块，给以后的泥浆搅拌带来不必要的麻烦。

3. Reading：Wall Plastering 1（墙面抹灰 1）

（1）Introduction

A good practice of plastering has something to do with the appearance of a building，it is important to apply plaster in the right order. Here the procedure of plastering will be discussed.

（2）Text

Once tools are cleaned，plaster is freshly mixed，and working surface and space are prepared well，it is time to do real plastering work on a wall. Here are the procedures for plastering.

1）Scoop a small amount of plaster out of a bucket and heap them on a hawk board.

2）Smear the plaster onto the wall，starting with the left bottom corner of the wall.

3）Crouch down and push the plaster up the wall in a gentle arc.

4）Slide the trowel over 2-3 inches at the top of the stroke，then reverse the motion and bring it down again.

5) Divide a wall into sections and plaster them from bottom to top, till a whole wall is plastered evenly.

6) Clear the trowel and run it over the wall in all directions to smooth the first coat of plaster.

7) Scrape the first coat of plaster to add some texture before applying the second coat.

8) Smear some plaster on the wall and apply the second coat in the same way as the first coat.

9) Glide the float slightly over the surface of the wet plaster in all directions to work out any lumps, holes, and inconsistencies.

In summary, what is explained above are some common practices, although they vary according to where they are conducted by whom. One thing in common is that attention shall be paid to each detail, such as focusing on the spot where the plaster is thicker.

(3) Words and Phrases

heap 堆积，装满

hawk board 镘板，带柄方形灰浆板

smear 涂抹，弄脏

crouch down 蹲伏，蜷伏

scrape 刮去，擦掉

float 镘刀，抹子

(1) 导读

泥浆涂抹的工序关系到房屋的外观质量，因此按照正确的步骤涂抹泥浆就显得十分重要。本篇将着重介绍泥浆涂抹的流程。

(2) 正文

一旦工具清洁完毕，泥浆刚刚搅拌完成，工作场地和墙面准备就绪，就可以开始给墙壁涂抹泥浆了。下面是涂抹泥浆的步骤：

1) 从泥浆桶中铲出少量的泥浆，然后将它们堆积在泥浆板上。

2) 从墙的左下角开始，将泥浆涂抹在墙上。

3) 蹲下身子将泥浆顺着一个小的弧度向上推。

4）在每一次推送泥浆快要结束时，把抹刀多划过 2～3 英寸，然后反方向再刮下来。

5）将整个墙体分成几个部分，从底部向顶部逐步地涂抹泥浆，直至整个墙体都涂抹平整。

6）将抹刀清洁干净后，用其在墙体上各个方向来回抹刮，使第一层的泥浆表面平滑。

7）在涂抹第二层的泥浆之前，在第一层的泥浆上用抹刀刮几下，以增加一些纹理。

8）在墙上再涂抹一些泥浆，参照涂抹第一道泥浆的方法进行第二道泥浆的涂抹。

9）在湿润的刚抹过的第二道泥浆表面上，用抹子轻轻地来回刮抹，表面的各个方向都需要轻刮一下，以便去除泥浆团块、空洞以及高低不一致的地方。

上面所讲到的是一些常规做法，而实际做法可能会有所不同，具体的步骤需要根据作业场所和施工人员来定。但是这里有一个共性，那就是都应该关注每一个细节，比如任何一个泥浆比较厚的地方。

4. Dialogue：Wall Plastering 2（墙面抹灰 2）

At 9 a. m. on Monday, on one apartment building construction site, Jack is learning the method of plastering from foreman John.

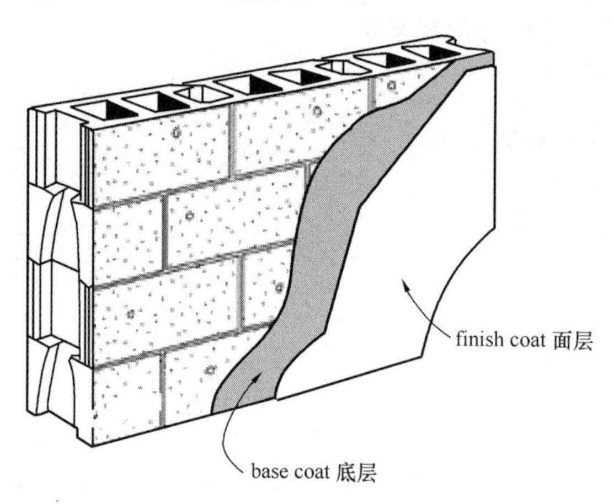

Fig. 9-1　Plastering 抹灰

John：Thorough preparation is key to achieving a professional finish, first, we start by covering the floor and any nearby furniture with protective sheets, and then remove all grease and dust from the surface.

Jack：How can we ensure a smooth even finish?

John：Apply plaster tape to cover all the joints between plasterboard sheets, use a sharp knife to trim the tape to a neat edge.

Jack：Is the next step to apply PVA glue?

John：Yes, roll the glue onto the wall. We start applying plaster once the glue is a little sticky to touch.

Jack: How to prepare the plaster mix?

John: Start by half filling a clean bucket with clean tap water, and then slowly add the plaster powder to the water.

Jack: What if there are some dry lumps of plaster around the edge of the bucket?

John: Use a bucket trowel to incorporate it into the mixture of plaster.

Jack: How are we going to mix them?

John: We do it by using a power mixer. Later on, we will clean the power stirrer.

Jack: Is it simple and easy to do with a power mixer?

John: Easy as it is, submerge it fully in the mixture, and set it at a low speed, then switch on, that is it. Remember never to overwork it.

Jack: How much plaster should be made now?

John: It's best to make some plaster mixture as and when we need it.

Jack: Are we ready for plastering now?

John: Yeah, we start by spreading the plaster firmly to the wall, using an upwards stroke with the trowel angled slightly away from the wall.

Jack: Where should we start from?

John: From the bottom left-hand corner of the wall, we work upwards and outwards, using a small amount of plaster each time.

Jack: How about those power sockets?

John: Disconnect the electricity before plastering. Of course, it's also best to cover the socket whilst working.

Jack: Shall we do plaster slowly?

John: Try our best to work quickly and build up a rhythm to cover the surface before the plaster begins to set.

Jack: Then second coat?

John: No, allow the base coat (as in Fig. 9-1) of plaster to dry slightly for about 20 minutes.

Jack: What is the next procedure?

John: We go over the plaster again with a trowel to smooth out any bumps, spraying some water on sections of the plaster several times.

Jack: Then we will proceed with the second coat?

John: Yeah, the same steps as for the base coat. The only difference is that the second coat needs to have a much smoother, more even, and a little thinner finish.

Jack: Is the plaster allowed to dry for a little while?

John: Yes, leave the plaster to dry until the surface is still slightly damp but firm enough that it doesn't move when touched.

Jack: Why do we spray water on the surface?

John: We dampen the surface and go over it lightly with a clean trowel blade.

Jack: Any tricks here?

John：Just angle the blade slightly，it will smooth the plaster and help to fill any small indentations.

Jack：What will we do next?

John：Use a small damp brush to polish off the corners and edges.

Jack：Are we done with our plastering yet?

John：No，let the plaster dry further before repeating this smoothing process for the final time.

Jack：Painting can be done after that smoothing?

John：Not immediately，the plaster should be left to dry completely before painting.

周一上午9点，在一个公寓楼的施工工地，杰克正在向施工队队长约翰学习抹灰工艺。

约翰：如果想要拥有一个光滑平整的墙面，那么全面和彻底的准备工作是关键。首先我们要把地板和所有的家具都用塑料布罩起来，然后把墙面上的浮灰和油渍清除掉。

杰克：怎样才能抹出一个平滑的墙面呢？

约翰：先用石膏绷带把所有的石膏板之间的接缝都粘结起来，然后用裁纸刀把石膏绷带的边沿裁齐。

杰克：下一步是滚涂PVA胶，是吗？

约翰：是的，把PVA胶在墙面上滚涂均匀。当PVA胶变得有些粘手的时候，就可以抹灰了。

杰克：怎样准备泥浆呢？

约翰：开始时，先在干净的水桶里盛放一半干净的自来水，然后一点一点地慢慢向其中加入石灰粉。

杰克：那如果水桶的边沿上有一些干的泥浆块，怎么办呢？

约翰：那就用抹刀把它们刮下来混进泥浆中。

杰克：我们要如何搅拌呢？

约翰：我们使用电动搅拌棒来搅拌，搅拌完成后要把搅拌棒清理干净。

杰克：搅拌棒使用起来容易吗？

约翰：非常容易，把电动搅拌棒全部浸没在泥浆搅和物中，将它设置为低速挡，打开电源开关即可。记住，搅拌时间不要过长。

杰克：现在我们要搅拌多少的泥浆？

约翰：最好是用多少就搅拌多少。

杰克：我们现在可以抹灰了吗?

约翰：可以了，开始时要用力把泥浆抹在墙上，要向上抹，注意抹刀要向外稍带些角度。

杰克：我们要从哪里开始抹灰呢?

约翰：从墙的左下角开始，向上向外抹，一次抹灰的量不要过多。

杰克：那些电源插座要如何处理呢?

约翰：在开始抹灰之前要把电源断开。当然，抹灰时最好能把它们给遮盖住。

杰克：我们抹灰是要慢一些吗?

约翰：不是的，抹灰要尽量抹快些，同时要保持一定的节奏。这样，在泥浆开始凝固之前，整个墙面的底层都能顺利地抹完。

杰克：然后是该抹第二遍的泥浆中层了吗?

约翰：还不行，底层（图 9-1）的泥浆需要干燥约 20min。

杰克：我们下一步该做什么呢?

约翰：我们要用抹刀在墙面上来回抹刮，试图抹平所有的凸起，这期间还需要时不时地在泥浆上喷些水。

杰克：下面是该抹中层的泥浆了吧?

约翰：是的。涂抹中层泥浆的方法和底层的基本一样，它们的不同之处是，中层泥浆需要更加平滑些，也要更薄一些。

杰克：还需要让中层的泥浆干燥一些时间吗?

约翰：需要的，要让它们干燥一些时间，直到中层还有一些潮湿，但用手触碰时泥浆不会再动。

杰克：我们为什么要喷水呢?

约翰：是想把表面湿润一下，以便用干净的抹刀刀刃轻轻地把它压抹平整。

杰克：这里有什么诀窍吗?

约翰：我们可以让抹刀刀刃与墙面稍微保持一个角度，然后轻轻地抹，这样能抹平一些小的凹坑。

杰克：我们接着要做什么呢?

约翰：用一个湿的小刷子来把边角刷干净。

杰克：到这里抹灰是不是就算结束了呢?

约翰：还不是，还需要让墙面再干燥一些时间，然后进行最后一遍的抹平。

杰克：最后一遍的抹平之后就可以刷漆了吗?

约翰：还不能立即刷漆，墙面需要完全干燥后才可以刷漆。

5. Exercises （练习）

(1) It is common to have dirt and __1__ on walls, all you need to do is use a sweeping

__2__ and sweep the walls clean from any dirt and dust.

(2) Mix the PVA in a large painting __3__ with 1-part PVA and 4-parts water, use a paint roller and paintbrush to __4__ the entire wall.

(3) Plug in the __5__ drill and put it into the flexible bucket with the bag of plaster without the mixer __6__ ; once the mixer is in the plaster, begin to increase the __7__ slowly and gradually, and move the mixer up and down trying to remove any __8__ , make sure to let the mixer get to the __9__ of the bucket.

Unit 10

Flooring and PVC Ceiling

（地板和吊顶）

1. Reading：Vinyl Flooring Installing （PVC 地板的安装）

（1） Introduction

Vinyl flooring is very typical in buildings nowadays, and easy to install. Here tools and procedures for vinyl flooring are discussed.

（2） Text

A vinyl plank flooring is durable, budget-friendly, and great for high moisture areas like basements and bathrooms. Its texture mimics the look of real hardwood. Another best part of it is that it is going to be one of the easiest flooring installations.

Normally, the tools used for vinyl plank flooring are a tape measure, hammer, jig-saw, undercut saw, and utility knife. Some of the most common materials are spacers and floor-leveling compounds.

When the tools and materials are ready, it is time to bring the required number of bo-xes of vinyl flooring plank into the room where they are acclimated to the room for 48 hours before use. It is better to use planks from different boxes to mix up colors and pat-terns.

What is next is to remove the baseboard, door jam, and others, if necessary. After that, prepare the subfloor as a clean, dry, and relatively level one, use a self-leveler on low spots and sandpaper on high spots.

Once floor preparation is done, score and cut the tongue off the first row, then set the first plank in place on the starting line, cut-side toward the wall. Hold the next piece at a slight angle and fold it down. To cut the last piece of the first row to fit, score with a util-ity knife and snap it. The end piece must be at least 6 inches, if this is not the case, cut a little bit off the first plank, and slide the row. That is all which is needed for the first row.

Now let us move on to row two. Insert the tongue of the first piece into the previous row and rotate down, for the next piece, first connect the short end, then the long end. Feel its lock. Keep doing this with the rest of the planks for row two, staggering the joints at least 6 inches and maintaining the expansion joint.

After finishing the second row, repeat the procedures of installation for row two till the whole flooring is completed.

To get under door jambs, the planks can be slightly bent into place, using a pull bar to lock the joint if needed.

To finish the installation, add transitions, replace the baseboard back into its original positions.

(3) Words and Phrases

texture 质感，质地
mimic 模仿，模拟
jigsaw 窄锯条机锯，往复式竖锯
score 划线于，刻痕于
staggered joint 交错式接缝，错缝
expansion joint 伸缩缝，胀缝

(1) 导读

如今，PVC 地板在房屋装修中应用得十分普遍，并且容易安装。本篇将着重介绍 PVC 地板铺设所需要使用的工具和其具体工序。

(2) 正文

PVC 地板耐用且价格实惠，适用于极端潮湿的空间，如地下室和卫生间。PVC 地板的质感和真正的硬木地板很接近，同时其另一个显著优点是铺设安装方法最简单。

通常来讲，铺设 PVC 地板所需要使用的工具包括卷尺、榔头、往复式竖锯、锯和文具刀，还有一些经常使用的材料，如垫片和地板修平粉等。

在材料和工具都准备齐全后，应把所需数量的 PVC 地板成箱地摆放在准备铺设地板的房间里，而且至少要提前 48h，以便 PVC 地板能够适应房间的温湿度状况。最好将不同箱子的楼板混合起来使用，这样颜色和花纹会比较协调。

如果需要的话，将踢脚线、门框等拆下。之后，准备一个干净、干燥且相对平整的地面，并在其低凹处使用自修平粉剂，在凸出处使用砂纸等工具进行打磨。

在地面准备好之后，将第一排所有用到的地板的榫舌用刀划掉，接着将第一块地板靠着起始线放下，使带切口的一面朝向墙。然后拿起第二块地板，与地面呈一个小的角度插

入后再折下去。第一排最后一块地板的尺寸应根据所剩下空间的大小来进行切割，可以使用文具刀划一下并掰掉多余的部分。最后一块地板应至少为 6 英寸长，若空间不够，则将第一块地板先切掉一些，然后再将整个第一排顺势滑过去。这样，第一排地板的铺设就算完成了。

下面将铺设第二排地板。将第二排地板的第一块板的舌插入第一排地板的槽里，之后向下按下。铺设第二块地板时，应先插入它的短边，然后再插入长边，要确认长、短两边都插入完成并啮合到位。按照同样的方法，将第二排剩余的地板铺设完毕，且两排地板之间的错缝应至少为 6 英寸，同时要注意预留伸缩缝。

在铺设完第二排地板之后，应按照同样的施工工序铺设其他排的地板，直到地板全部铺设完毕。

在门框下面的部分，可以将地板轻轻弯曲一下再插进去。如果有必要的话，可以使用撬板将地板插入对应的槽里。

在完成地板的铺设之前，还需要加装过渡条，并重新安装踢脚线。

2. Reading：Tile Flooring（瓷砖地板）

(1) Introduction

To have a beautiful tile flooring, it is necessary to mix the thinset mortar and grout thoroughly, cut the tile and membrane in a good size, place the tiles and press the grout into the joints in a proper way. Here the tools and methods of laying a tile flooring will be discussed.

(2) Text

A tile floor installation creates a beautiful, durable floor which is easy to clean. It is common to be seen in the bathroom, kitchen, and living room of a modern house.

Generally speaking, the tools used for tile flooring are straight edge, notched trowel, pointing trowel, margin trowel, flush-cut saw, tape measure, tile cutter, nipper, wet saw, level, and rubber mallet.

Before laying tiles, tile membrane, waterproof membrane tape, floor tile, tile spacer, and thinset mortar should be ready and handy. Thinset mortar is the adhesive that will hold the tile to the floor.

Once materials and tools are prepared well, tile flooring can be commenced, here are steps on how to lay tiles:

1) Prepare the Subfloor

The most important step for a tile installation is preparing the subfloor properly, the main thing to look for is a level and even subfloor, if there are any low spots, fill them with a leveling compound. Another thing is to plug the sewer pipe with a rag.

2) Cut the Tile Membrane to Size

Roll out the membrane with the fleece side down, and then mark the subfloor at the edge of the membrane, that is where to spread the thinset to. If any pipes are installed on the floor, cut the membrane around the pipes using a utility knife.

3) Spread and Comb the Thinset for the Tile Membrane

① Mix the thinset, and put the trowel into it to see whether or not the ridges can stay standing up, if yes, the mixture has the right consistency and is ready for use.

② Work one section at a time so the thinset would not dry before installing the membrane.

③ Spread the thinset onto the subfloor with the smooth side of the trowel, covering the whole section evenly. Then use the notched side of the trowel to comb the mortar.

④ Roll out the membrane and press it down into the thinset.

⑤ Keep doing in sections, applying mortar and then membrane.

⑥ Spread some thinset onto the membrane and embed the waterproofing tape into it with the trowel, at least a 2-inch overlap on each seam shall be allowed.

4) Create a Starting Point for Laying Tile

Measure two opposite walls, then snap a chalk line between the two center points, next do the same for the remaining two walls, therefore, reference lines are marked.

5) Dry Layout

Dry fit the tiles to check the layout, use tile rubber spacers to make even joints, and leave a 1/4-inch gap along the outside edges for expansion. Mix tiles from different boxes to have a color consistency for the whole flooring.

6) Spread the Thinset before Laying the Tile

Mix some more thinset thoroughly till a peanut butter consistency is attained. Spread the thinset evenly from the center, filling the cavities in the membrane.

Work a section at a time so that the thinset can be kept from drying out before laying the tiles. Comb the thinset with the notched side of the trowel.

7) Lay Tile Flooring

① Gently lay a tile on thinset along the reference line in the center of the room, with fingertips widespread, push down with a slight twist of the wrist.

② Place rubber spacers between each tile.

③ Keep doing the same with the tiles for the rest of the row.

④ Continue spreading thinset and setting tiles in 2-by-3 foot sections, working from the center of the room out toward the walls, till the whole flooring is done.

⑤ Use a trowel to scrape off any thinset from the tile surface or in the joints.

8) Fill the Joints with Grout

① After laying the whole tile floor, let the thinset dry for 24 hours before grouting.

② Remove the spacers between the tiles.

③ Mix up a batch of grout thoroughly by adding water a little at a time.

④ Scoop a trowel of grout onto the floor.

⑤ Press the grout into the joints with a rubber float and pull the grout across the joints diagonally. Remove the excess.

⑥ Allow the grout to set up for 20 minutes and then wipe the grout lines with a wet sponge and clean water.

⑦ Once the grout is in, wait 72 hours and then use a grout haze remover to remove any haze left on the surface.

⑧ Wipe with a clean sponge, and repeat wiping until the tile is clean.

(3) Words and Phrases

straight edge 直尺，平尺

notched 有切口的，带凹口的

pointing trowel 勾缝镘刀，尖形抹子

margin trowel 抹边镘刀，边角抹子

flush cut saw 平头锯

tape measure 卷尺

tile cutter 砖切割工具

nipper 夹石钳

wet saw 湿式切削瓷砖锯

level 水平尺

rubber mallet 皮锤

membrane 薄膜

rag 破布

fleece 起绒的

ridge 刃，边缘

chalk line 粉笔线

cavity 空心，凹处

grout haze 砂浆残留

(1) 导读

如果想要铺贴一个漂亮的瓷砖地面，那么就必须保证瓷砖胶浆和灰浆搅拌均匀，瓷砖和薄膜的切割尺寸合适，且瓷砖铺贴的方法和勾缝工序符合规范要求。本篇将着重介绍瓷砖铺贴所需要的工具和其具体工序。

(2) 正文

瓷砖地面漂亮耐用，且容易清洁，在如今的房屋建筑中经常被用于卫生间、厨房和起居室。

通常来讲，铺贴瓷砖需要使用的工具有：直尺、齿形抹刀、勾缝抹刀、抹边抹刀、平头锯、卷尺、砖切割工具、夹石钳、湿式切削瓷砖锯、水平尺和皮锤。

在铺贴瓷砖之前，应该将薄膜、防水薄膜带、瓷砖、瓷砖十字胶粒和瓷砖胶浆等准备就绪。瓷砖胶浆是一种胶粘剂，用来将瓷砖和地面粘结在一起。

一旦材料和工具准备完成，就可以正式铺贴瓷砖了。下面是铺贴瓷砖的具体步骤：

1) 准备地面

瓷砖铺贴工序中最重要的一项准备步骤是保证地面平整，如果有任何低凹的地方，应用找平剂将它们填平。此外，应用破布将污水管道口堵住。

2) 裁切大小合适的瓷砖薄膜

打开成卷的薄膜，使其起绒面朝下并将它铺开，然后沿着地面的外沿在薄膜上相应地作标记，标注之后瓷砖胶浆需要抹涂的地方。如果地面上安装有任何管道，可以用剪纸刀沿着管道来切割薄膜。

3) 对瓷砖薄膜抹刮瓷砖胶浆

① 在搅拌瓷砖胶之后，将抹刀放进胶中，并观察抹刀的刀刃能否直立，若能够直立，则表明瓷砖胶的黏稠度是适宜的。

② 每次仅在一个区域进行抹刮，这样可以保证在铺设薄膜之前瓷砖胶不会干燥。

③ 用齿形抹刀的直边将瓷砖胶抹刮在地面上，整个区域都需要抹刮平整，然后用抹刀的齿形边用力刮出泪状条纹。

④ 把薄膜圈打开、摊开后，将其按压在瓷砖胶上。

⑤ 一个区域接着一个区域地进行，先抹胶后铺膜。

⑥ 在薄膜上抹刮一些瓷砖胶，用抹刀将防水带压入瓷砖胶中，且每个接缝至少要有 2 英寸的搭接长度。

4）弹十字线

对两面对立的墙分别进行测量，然后在两面墙的中心点之间弹出一条线，在另外两面墙之间也拉出一条中心线，这样就为瓷砖的铺贴准备好了十字参考线。

5）预铺

随后进行瓷砖的预铺，检查一下整体的效果。在接缝中放入十字胶粒来确保接缝的宽度均匀一致，在外沿边预留四分之一英寸的缝隙作为伸缩缝。将不同箱子中的瓷砖混合起来使用，这样可以让整个地板的颜色均匀一些。

6）在铺贴瓷砖前抹涂瓷砖胶浆

充分搅拌一定量的瓷砖胶浆，直至其稠度类似花生酱。从地面的中心处向外均匀地抹涂瓷砖胶浆，注意要填满薄膜的空隙处。

每次只抹刮一个区域，这样可以保证在铺贴瓷砖前，瓷砖胶浆不会干燥。然后，用抹刀的齿形边在瓷砖胶浆上用力刮出泪状条纹。

7）铺贴瓷砖

① 在地面的中心处沿着参考线将瓷砖轻轻地放置在瓷砖胶浆上，手指摊开，用手腕轻轻地旋转来按压瓷砖。

② 在瓷砖之间放上橡胶十字胶粒。

③ 用同样的操作方法将同一行的其他瓷砖铺贴完毕。

④ 每次取 2×3 英尺的一个区域，继续抹涂瓷砖胶浆并铺贴瓷砖，从地面的中心开始，向外、向墙的方向逐步进行，直至整个地面铺贴完毕。

⑤ 用抹刀刮掉任何残留在瓷砖表面或接缝里的瓷砖胶浆。

8）勾缝

① 在整个地面的瓷砖铺贴完毕之后，要保证瓷砖胶浆干燥 24h，才能进行勾缝。

② 首先，取出瓷砖之间的十字胶粒。

③ 彻底、均匀地搅拌灰粉，每次加少量的水进行搅拌。

④ 用抹刀铲出一抹刀的灰浆到瓷砖上。

⑤ 用橡胶抹刀将灰浆推压进各个瓷砖缝隙里，再按相反方向、与瓷砖缝隙成对角线的方向刮回，并及时刮掉多余的灰浆。

⑥ 等待灰浆凝结 20min，然后在清水里蘸湿海绵，并用它来清洁瓷砖缝。

⑦ 勾缝 72h 之后，用灰浆清洁去除剂清除瓷砖上的任何残留胶浆。

⑧ 反复用湿海绵清洁瓷砖地板，直至瓷砖地板干净。

3. Reading：PVC Ceiling Installing （PVC 吊顶安装）

（1） Introduction

PVC ceiling panel is popular for both residential and industrial buildings for its easy installation and durability, how to fix a PVC ceiling panel will be discussed here.

(2) Text

Installing a PVC ceiling in the kitchen, bathroom, and living room has such benefits as durability and affordability, for both domestic and commercial spaces.

Tools needed to install PVC ceiling include a tape measure, saw, workbench, Stanley knife, pencil, silicon gun, and drill with a hole saw.

When the tools and materials are ready, it is time to commence the installation of the ceiling. How to do it is as follows.

1) Preparation

Batten out the ceiling, ensuring it is completely level.

2) Cutting the Edging Trims

Measure the ceiling perimeter walls and cut the edging trims to the length.

3) Fixing the Edging Trims

Fix the edging trims to the perimeter of the ceiling, using screws approximately 300mm apart, along three sides of the room, overlapping into the corners of the room.

4) Cutting the First Panel

① Measure the ceiling wall and transfer the measurement onto the first panel with a straight pencil line.

② Catch the first panel and cut it to size with a saw, keeping the saw-blade shallow on each stroke to provide a smooth finish.

③ Cut off the male tongue of this first panel along the length.

④ Insert both two short ends of the first panel into the pockets of edging trims and then push the long edge of the panel home into the edging trim, thus, the first panel is fitted into place on the ceiling and secured.

⑤ Cut the second panel to length after measuring and slot it into the first one with the tongue fitting into the groove of the first panel.

⑥ Repeat the same procedure for all the remaining panels, locking each one into place with the tongue and groove edge.

5) Installing the Final Panel

① Cut the final panel down to suit the width of the gap.

② Disperse a good-quality grip adhesive on it.

③ Attach the final edging trim to the final panel.

④ Press the final panel into place carefully, ensuring each end is tucked into the side trim securely.

One more thing is about the light's installation on the ceiling panel. When installing the LED lights, first make some marks on the ceiling panel for them, and then put the drill with a hole saw at low speed, next drill the hole in the ceiling, finally install the lights on the ceiling.

(3) Words and Phrases

batten out 板条，木条
dispense 分配，配发
edging 边缘，边框
grip 紧握，紧咬，
groove 沟，企口，槽
overlapping 搭接，重叠
silicon gun 有机硅枪
slot 把……放入槽内
tongue 雄榫，榫舌
trim 镶边修饰
work beach 工作台

(1) 导读

PVC 吊顶在工业和民用建筑中经常使用，其非常耐用，且安装简单方便。本篇将着重介绍 PVC 吊顶的安装方法。

(2) 正文

在厨房、卫生间和起居室里安装 PVC 吊顶有诸多优点，比如耐用、价格合适。它适用于工业和民用建筑。

安装 PVC 吊顶所需的工具包括卷尺、锯、工作台、文具刀、铅笔、硅胶枪和打孔钻。

在工具和材料准备齐全之后，就可以开始安装吊顶了。下面是安装 PVC 吊顶的具体步骤。

1) 准备

在准备安装吊顶的屋顶上找到适当位置钉上木条，同时确保木条处于一个水平面。

2) 截取用作边缘装饰的硬质卡槽

对准备安装吊顶标高处的周围墙壁进行测量，并根据实际测量值来截取相应长度的硬质卡槽。

3）安装硬质卡槽

将用作边缘装饰的硬质卡槽钉在三面墙的吊顶标高处，使其与房屋墙角重叠，同时保证两个钉的间距为300mm。

4）切割第一块吊顶板

① 测量吊顶标高处墙的宽度，并根据实际测量值，在第一块吊顶板上用铅笔画上一条直线。

② 取第一块吊顶板，沿着上述铅笔线锯开，每次锯切时锯齿不要太深，这样可以保证锯切面的平滑。

③ 将第一块吊顶板的榫舌全部切割掉。

④ 将第一块吊顶板的两个短边分别插入两个硬质卡槽的空槽里，然后将其一长边推进另外一个硬质卡槽的空槽里，这样第一块吊顶板就安装到位了，然后再将其固定好。

⑤ 根据实际测量值截取第二块吊顶板，并将其插入第一块吊顶板里，让其榫舌插入第一块吊顶板的企口里。

⑥ 按照同样的方法截取并安装剩余的吊顶板，需要保证它们的榫舌与企口相互啮合。

5）安装最后一块吊顶板

① 最后一块吊顶板的宽度应根据实际剩下的空隙来确定和截取。

② 在吊顶板上挤涂高质量、粘结力强的胶粘剂。

③ 将最后一条用作边缘装饰的硬质卡槽粘到末板上。

④ 将最后一块吊顶板小心仔细地压入相应位置，确保每个端边都能可靠地插入边框里。

此外，还要解决吊顶上安装灯具的问题。在安装LED灯时，首先要在吊顶板的相应位置做好标记，将带开孔头的电钻设置为低速挡，然后在吊顶上打孔，最后将LED灯安装在吊顶上。

4. Dialogue：Procedures of Installing Vinyl Flooring（PVC地板的安装工序）

At 10 a.m. on Monday, at one apartment building site, two workers were installing Vinyl flooring in one house, Jack took David there to have a look around.

David：There are some patches with different colors, here in the living room, why is that?

Jack：Since there were several low-lying areas on the concrete subfloor, they spread some small trowels of patching compound to fill last week, to have a flat and smooth subfloor.

David：Why do they make marks on the Vinyl Flooring over there?

Jack：Oh，those marks say where the excess will be cut out. If there are some obstructions，they make relief cuts.

David：What do they apply double-stick tape for?

Jack：That tape creates a bond between vinyl backing and the template made of craft paper.

David：I see，they are cutting the vinyl along the template edge. What is next?

Jack：They will open a can of adhesive materials，and spread a band along the wall about 6 inches wide.

David：They would roll the vinyl flooring back down into the position，seat the flooring with a hand roller?

Jack：Yes，they will leave the finished flooring there，allowing it to dry for two to four hours.

David：When will they replace shoe molding and trim?

Jack：That is the last step，just when there is a solid bond between vinyl flooring and subfloor.

David：Oh，it is almost lunch time.

Jack：Alright，let us go back to the office. We will come back here tomorrow to have another look.

周一上午 10 点，在一个公寓楼的施工现场，几位工人正在一间房屋里安装 PVC 地板，杰克正带着戴维在现场检查巡视。

戴维：起居室有几个水泥补丁，它们的颜色不同，为什么会这样呢?

杰克：水泥楼面底板上原本有几个凹坑，他们上周用了一些修补材料来填平，这样就可以使楼面底板平整、平滑。

戴维：那边他们在地板上做了一些标记，这又是为什么呢?

杰克：他们会在有标记的地方把多余的 PVC 地板裁切。如果地板上有一些需要避开的地方，他们也会在相应的位置做上一个切口。

戴维：双面胶是用来做什么呢?

杰克：双面胶用来把 PVC 地板的背面和牛皮纸模板粘结起来。

戴维：明白了，他们正沿着模板的边沿裁切 PVC 地板。他们下一步会做什么呢?

杰克：他们会打开一罐胶，然后沿着墙涂敷大约 6 英寸宽的胶粘带。

戴维：然后他们再把 PVC 地板重新滚回原来的位置上，用一个辊子在其上来回滚，使 PVC 地板与胶能有效地粘结起来，是吧?

杰克：是的，他们会保证铺好的PVC地板和胶有2～4h的干燥时间。

戴维：他们会在什么时间重新安装门框和踢脚线呢？

杰克：那是最后一步了，需要等到PVC地板与楼面底板充分地粘结完成之后。

戴维：已经快到午餐时间了。

杰克：好的，我们现在就回办公室，明天再过来检查一下。

5. Exercises （练习）

（1） Start by removing the baseboards. First, cut through any paint sealing the baseboards to the wall using a __1__. Then use a __2__ to gently loosen the baseboards.

（2） Score and cut the __3__ off the first row of vinyl planks using a utility knife. Set the first plank in place on the __4__ line with the cut side toward the wall, maintaining the expansion gap.

（3） Insert the tongue of the first piece into the __5__ of the previous row's first plank and rotate __6__ to click together.

Key to Exercises (答案)

Unit 1

1. aggregate 2. compressive 3. softwoods 4. layers 5. carbon

Unit 2

1. table 2. cave-in 3. lines 4. compacting 5. heavy equipment

Unit 3

1. piles 2. type 3. life 4. table

Unit 4

1. fall 2. access 3. safe 4. Unauthorized 5. structure

Unit 5

1. bond 2. fit 3. hold 4. rebars 5. intervals 6. reinforcing 7. configuration 8. plans 9. position

Unit 6

1. sample 2. batch 3. quantity 4. frequency 5. scooped 6. coated 7. voids 8. curing 9. compression 10. specimens

Unit 7

1. cement 2. sands 3. hardens 4. dry 5. mixer 6. additive 7. gaps 8. binding 9. retaining 10. attach

Unit 8

1. layers 2. repellent 3. sheets 4. entry 5. moisture 6. film 7. compounds 8. mixing 9. walls 10. cavity

Unit 9

1. dust 2. brush 3. tray 4. cover 5. mixing 6. spinning 7. speed 8. lumps 9. bottom

Unit 10

1. utility knife 2. pry bar 3. tongue 4. starting 5. groove 6. down

参 考 文 献

[1] 周明鑑，魏向清．综合英汉科技大词典(第2版)[M]. 北京：商务印书馆，2016.
[2] 朱从明．建筑施工技术[M]. 北京：航空工业出版社，2015.
[3] 钟汉华，张天俊．建筑施工技术[M]. 北京：人民邮电出版社，2015.
[4] S. S. Bhavikatti. Basic Civil Engineering[M]. New Delhi：New Age International Publishers，2010.
[5] 《土木建筑工程英文词典》编写组．土木建筑工程英文词典[M]. 北京：中国水利水电出版社，2008.